GW01402987

few days

love lyn X.
7/6/07

THE REED FIELD GUIDE TO
NEW ZEALAND
ALPINE
FAUNA AND FLORA

For Tim Wilson and Folau Lea'aetoa

Something missing go and find it
Go and look beyond the ranges;
Something lost beyond the ranges,
Lost and waiting for you. Go.

Rudyard Kipling

THE REED FIELD GUIDE TO

NEW ZEALAND

ALPINE

FAUNA AND FLORA

Brian Parkinson

REED

KEY TO PHOTOGRAPHERS AS INDICATED BY INITIALS BESIDE PHOTOGRAPHS

JB	John Braggins	CM	C.H. Monte	BP	Brian Patrick
RB	Robin Bush	GM	Geoff Moon	NS	N.C. Simpson
PG	Peter de Graaf	RM	Rod Morris	IS	Ian Southey
RMM	R.M. McDowall	SN	Steve Newall	RS	Rob Suisted

FRONT COVER: Kea (RM)
BACK COVER: Alpine Black Caddis (BP), Lupin and Broom (CM), Superb Land Snail (RM)
TITLE PAGE: White Sun Orchid (RM)
CONTENTS PAGE: Alpine Green Stoner (BP)

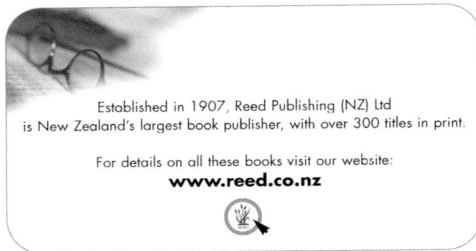

Established in 1907, Reed Publishing (NZ) Ltd
is New Zealand's largest book publisher, with over 300 titles in print.

For details on all these books visit our website:
www.reed.co.nz

Published by Reed Books, a division of Reed Publishing (NZ) Ltd,
39 Rawene Rd, Birkenhead, Auckland 10.
Associated companies, branches and representatives throughout the world.

ISBN 0 7900 0785 1
First published 2001

Edited by Carolyn Lagahetau
Designed by Graeme Leather
Scanning and Prepress by i2i imaging
Printed in Singapore

Contents

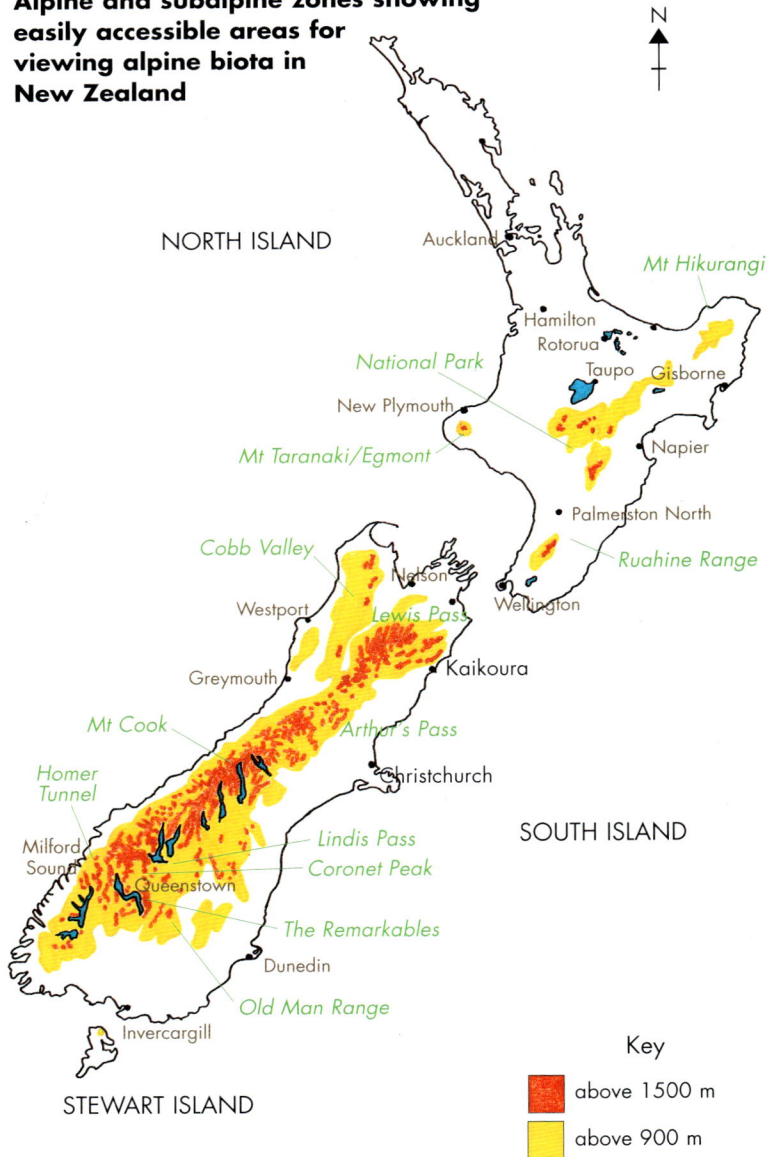

Alpine and subalpine zones showing easily accessible areas for viewing alpine biota in New Zealand

N

NORTH ISLAND

Auckland

Mt Hikurangi

Hamilton
Rotorua

National Park

Taupo   Gisborne

New Plymouth

Mt Taranaki/Egmont

Napier

Palmerston North

Cobb Valley

Ruahine Range

Nelson

Lewis Pass

Westport

Wellington

Greymouth

Kaikoura

Mt Cook

Arthur's Pass

Homer Tunnel

Christchurch

Milford Sound

Lindis Pass

SOUTH ISLAND

Coronet Peak

Queenstown

The Remarkables

Dunedin

Old Man Range

Invercargill

STEWART ISLAND

Key

above 1500 m

above 900 m

6

Introduction

In New Zealand, 'alpine' is a generic term covering all of the terrain above the timberline. This is the territory covered by this book.

The biota

The newness of our mountains, following a long period of quiescence when much of New Zealand was a low and flat peneplain, means that the plants and animals that live here have made the necessary adaptations to live in these rigorous and seemingly inhospitable conditions remarkably quickly.

These groups of plants and animals either consist of lowland forms that were forced to adapt to an alpine existence, or they are migrants that have island-hopped to New Zealand from the alpine areas of Australia and elsewhere.

After the emergence of the mountains, other factors that were crucial in shaping the evolution of the alpine biota were the many periods of glaciation and inter-glaciation that occurred during the last Ice Age. During this time, vast sheets of ice covered most of the South Island and extended probably as far north as the Tararua Ranges in the lower North Island, eliminating virtually all life in their path.

It is thought that the permanent snowline during these periods was at least 1000 m lower than it is at present, and that tussock and subalpine scrub replaced forest over a large part of the South Island. Animals that were unable to escape northwards were forced to learn to live in this new environment.

These factors created pockets where the conditions were not so harsh, enabling some plants and animals to survive. These pockets, known as *refugia*, resulted in many discrete populations being formed, thus creating optimum conditions for speciation.

One of the interesting results of the glaciations was that a number of species were eliminated over a large part of their range in the South Island, leaving remnant populations at opposite ends of the South Island in Fiordland and Nelson where, for various reasons, conditions were not so severe. Even though the last Ice Age finished some 20,000 years ago, these isolated species have still not been able to recolonise the intervening areas.

The mountains

The key to understanding the mountain-building in the South Island of New Zealand is to remember that the motive power is provided by two huge, slow-moving continental plates, the Pacific Plate and the Indo-Australian Plate. These plates are being crushed together, squeezing mountains upwards, a little like toothpaste coming out of a tube.

Where the pressures are greatest the rocks reform as the hard and shiny schists of Otago, and where the rocks are cracked water provides an inexorable splitting power as it expands while forming ice. The most dramatic sign that massive natural forces are at work is in the great faulting gash that marks the regions of collision which extends the length of the South Island.

Although New Zealand first separated from Australia to become a distinct entity around 100 million years ago during the Cretaceous period, the mountains of New Zealand are, in geological terms, relatively young. They resulted from the Kaikoura orogeny (or mountain-building phase) that began at the end of the Oligocene epoch and continued into the early Miocene epoch, some 25 million years ago.

It was not until the end of the Miocene epoch, some 7 million years ago, that the first high country began to appear, with the first terrain that could truly be regarded as alpine emerging only around 2 million years ago. Since then, further growth has been kept in check by the eroding influences of rain, wind and freeze-thaw action, and an uneasy equilibrium is thus maintained.

Although activity was strongest along a line running on a northeast to southwest axis from East Cape through to Fiordland, this activity occurred at differing intensitites at different places at disparate times. The Kaikoura Ranges were the first to appear, followed by the Southern Alps, which have been uplifting along the Alpine Fault since the Pliocene. The mountains of the North Island are even younger, as they have been around only since the late Pliocene. The island's distinctive, high volcanic mountains have appeared even more recently, not much more than a million years ago.

These 'ups' and 'downs' are still going on. Mountains like the Southern Alps are still emerging but are just as rapidly being eroded away.

One particularly interesting aspect of the New Zealand mountains is the presence of screes. These are the result of the fracturing of the friable rocks found in the greywacke mountains that extend from Marlborough to North Otago. The resulting rubble cascades downwards, propelled by gravity and a process known as solifluction, forming characteristically unstable habitats to which a number of truly

Time chart

million years ago

		Recent		Ice ages
2	**Cenzoic**	Pleistocene		
7			Pliocene	Kaikoura Orogeny
25		TERTIARY	Miocene	
37			Oligocene	
54			Eocene	
65			Palaeocene	
	Mesozoic	CRETACEOUS		Opening of Tasman Sea First angiosperms
135				Rangitata Orogeny
		JURASSIC		
190				
		TRIASSIC		New Zealand Geosyncline
230				
	Palaeozoic	PERMIAN		
280				
		CARBONIFEROUS		
350				
		DEVONIAN		
390				Tuhua Orogeny
		SILURIAN		
440				
		ORDOVICIAN		Buller Geosyncline
500				
		CAMBRIAN		
570				
		PRECAMBRIAN		

remarkable animals and plants have adapted. Twenty-five alpine plants from a number of quite different families are found on screes, along with a lizard and a number of insects and spiders.

The zones

It soon becomes apparent to the observer that the vegetation on mountains is confined to distinct zones. As one climbs higher, one encounters distinctive and predictable assemblies of plants that are generally associated with particular areas, governed by climate and conditions.

Each of these zones extends over a range of 300–500 metres. Although one soon learns what sort of vegetation to expect at any particular altitude, there are many variables that need to be taken into account. Among the most significant of these are the direction the particular slope faces: does it face north (warmer), south (cooler), east (drier) or west (wetter)? For example, plants can grow at higher altitudes on warmer, north-facing areas than they can on cooler, south-facing areas.

As one proceeds southwards in New Zealand into the higher latitudes and as the climate consequently becomes colder, these zones get lower. The upper limit of the treeline or timberline is defined by Hugh Wilson, the New Zealand alpine plant authority, as the upper limit of erect, medium to large trees. On the Volcanic Plateau of the central North Island, the northernmost of our alpine areas, this treeline is at about 1450 m, whereas it is at an altitude of approximately 900 m in places like the Tin Range on Stewart Island and in the Takitimu Mountains of Southland which lie in the far south of mainland New Zealand.

**Geological zones
of New Zealand**

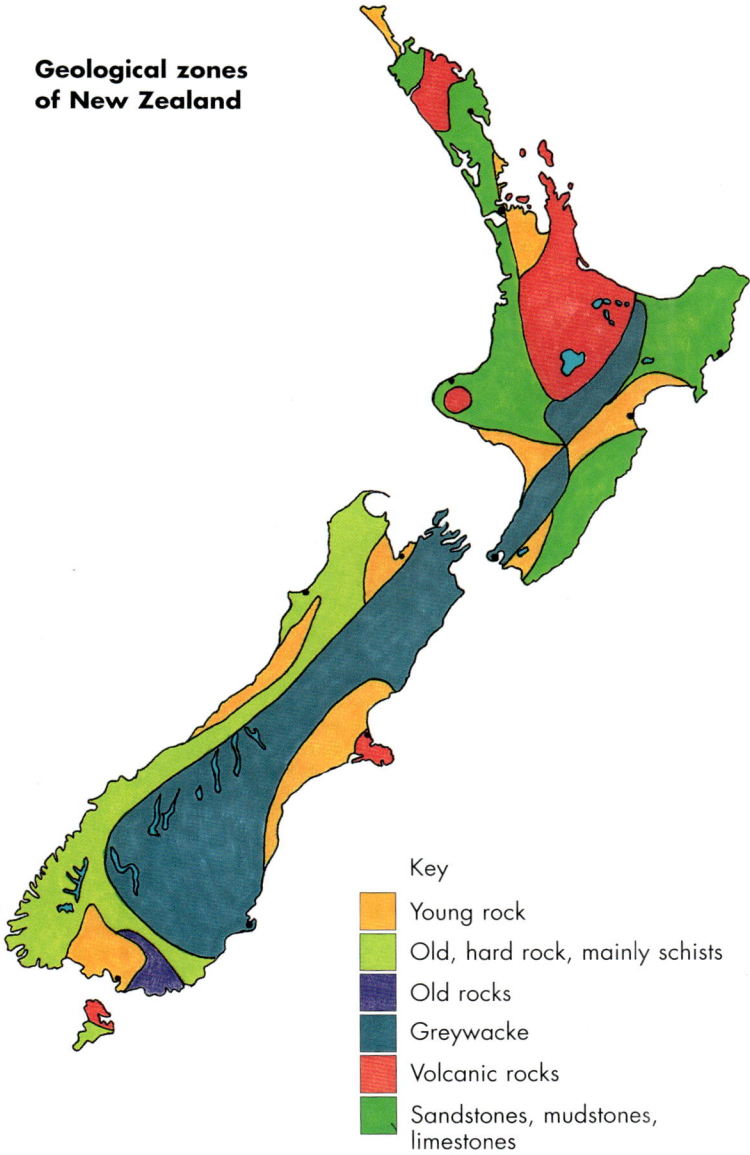

Key

- Young rock
- Old, hard rock, mainly schists
- Old rocks
- Greywacke
- Volcanic rocks
- Sandstones, mudstones, limestones

1 Nival: > 2000 m.
Snowfields, bare rocks, moist zones and debris.

2 High alpine: 1500–2000 m.
Tussocklands, bare rock and herbfields.

3 Low alpine: 1200–1500 m.
Screes and moraines.

4 Subalpine: 800–1200 m.
Scrub, shrubland, tussock grasslands, bare rock, screes, moraines and riverbeds.

5 Montane: 450–800 m.
Forest.

Mammals

Mammals present interesting anomalies in our alpine biota in that, unless native bats occasionally turn up here at these higher altitudes, all mammals are exotic and were either deliberately or accidentally introduced to New Zealand.

Only Chamois and Tahr were deliberately released into the alpine areas but several other mammal species have migrated upwards from lower altitudes. Without exception they stop only on the tops during the summer, descending to the shelter of scrub or beech forest during colder periods.

Browsing mammals cause significant damage to alpine flora and are a major cause of erosion in the less stable mountain areas. They are also implicated in the decline of birds like the Takahe, either by competing for food sources, as in the case of the Red Deer, or by predation by animals such as Stoats. Intensive control by government and private hunters has now eliminated these animals from many areas, allowing the alpine flora to recover some of its lost ground.

Himalayan Tahr   *Hemitragus jemlahicus*

CHARACTERISTICS	Height at shoulder — 1 m.
BIOLOGY	Herbivore. Ruts in May and June. Breeds during spring.
DISTRIBUTION	Alpine areas of the central South Island, from about Arthur's Pass, south to around Lake Wanaka.
HABITAT	Steep tussock-covered slopes between 1300 and 2000 m. Descends to more sheltered areas during blizzards.
NOTES	Originally from the Himalayas, the Tahr in New Zealand are descended from animals introduced from England. They cause severe damage to alpine plants by grazing. The present population is estimated at about 15,000.

Chamois   *Rupricapra rupicapra*

CHARACTERISTICS	Height at shoulder — 80 cm.
BIOLOGY	Herbivore. Ruts in May and June. Breeds during spring.
DISTRIBUTION	Widely distributed throughout alpine and subalpine areas of the South Island, from the Seaward Kaikouras south to about Lake Wakatipu.
HABITAT	Areas of scrub, scree and tussock. In snowy periods, chamois descend to lower, more sheltered areas.
NOTES	Introduced from Europe in the early 1900s. The present population is estimated at around 30,000 animals.

RM

Himalayan tahr — mature bull.

RS

Chamois

Red Deer *Cervus elaphus scoticus*

CHARACTERISTICS	Height at shoulder — 1.1–1.3 m.
BIOLOGY	Breeds in spring. Stags bellow and roar during the rut in May and June.
DISTRIBUTION	Widely distributed throughout the North and South islands from sea level to moderate altitudes. In summer they sometimes range upwards into alpine scrub and grasslands.
HABITAT	Scrub, grasslands and forested areas.
NOTES	An introduced species that damages native flora through grazing. Interbreeds with the related Wapiti where their ranges overlap.

Wapiti *Cervus elaphus nelsoni*

CHARACTERISTICS	Height at shoulder — 1.2–1.5 m.
BIOLOGY	Herbivore. Breeds in spring. Stags bellow and roar during May and June.
DISTRIBUTION	In Fiordland, from Martin's Bay south to about Doubtful Sound. In summer sometimes ranges upwards into alpine scrub and grassland.
HABITAT	Scrub, grasslands and forested areas.
NOTES	Introduced from North America in the early 1900s. The largest deer species found in New Zealand.

Red Deer — roaring stag.

Wapiti bull.

Feral Goat *Capra hircus*

CHARACTERISTICS	Height at shoulder — 50–60 cm.
BIOLOGY	Herbivore. Generally mates in autumn and gives birth in spring but sometimes also breeds at other times of the year.
DISTRIBUTION	Widespread in the North and South islands, from sea level up into alpine areas. Retreats into sheltered, forested areas in bad weather.
HABITAT	Forest, scrub and grasslands.
NOTES	First introduced to New Zealand by Captain James Cook in 1777 from England. Causes serious damage to native forest and alpine plants by indiscriminate grazing.

Bennett's Wallaby *Macropus rufogriseus*

CHARACTERISTICS	Height at shoulder — 60 cm.
BIOLOGY	Herbivore. Breeds during spring, usually giving birth to a single young called a 'joey'. Females mate soon after giving birth, but the embryo remains dormant until the pouch is vacated.
DISTRIBUTION	Widespread in South Canterbury. Another smaller population in Central Otago.
HABITAT	Along margins of forest, in scrub and in tussocklands.
NOTES	Introduced from Australia, this is the most widespread of our five wallaby species. A pest in forests and on pastureland.

Feral Goat — mature male.

Bennett's Wallaby

Stoat *Mustela erminea*

CHARACTERISTICS	Length — 35–40 cm.
BIOLOGY	Carnivorous. Takes only live prey such as birds, rodents and lizards.
DISTRIBUTION	Widespread in the North and South islands from sea level to moderate altitudes. Sometimes ranges into alpine areas in summer.
HABITAT	Grasslands, scrub and forested areas.
NOTES	Introduced from Britain in the nineteenth century to control rabbits. A serious predator of native wildlife, and a major contributor to the decline and extinction of many bird and lizard species.

Hare *Lepus europaeus*

CHARACTERISTICS	Length — 60–70 cm.
BIOLOGY	Herbivorous, eating mostly shoots and grasses. Males chase and box one another in spring. Mating takes place in the winter and the young, called leverets, are born in late spring and summer.
DISTRIBUTION	Widely distributed in the alpine areas of both North and South islands.
HABITAT	Areas of scrub and native grasslands. Sometimes along the margins of pine forest. Spends the daylight hours in a small hollow or 'form' in the grass.
NOTES	Introduced from England in the nineteenth century for hunting. Considered a pest in some farming areas and in some pine forests.

Stoat

Hare — young female.

Birds

Although birds are relatively plentiful in the mountains of New Zealand, particularly during the summer months, there is probably only one bird species that can be considered truly alpine, remaining at high altitudes all year round — the Rock Wren. Even that hardy character, the Kea, moves to lower altitudes during the colder months if the weather turns really foul.

When spring returns, birds like the Pipit move up into the subalpine and alpine zones and during the summer months they are regularly encountered breeding here, retreating to the lowlands again at the onset of colder weather. Others, like the Harrier Hawk, are often seen quartering the terrain in search of prey. Species like the Black-backed Gull and the Black-fronted Tern, which one would normally expect to find at the seaside, are regular visitors to the upper stretches of braided rivers and tarns.

Other birds like the Black Shag, *Phalacrocorax carbo*, and the Grey Duck, *Anas superciliosa*, are regularly seen in alpine areas. They are usually in transit between their various lowland haunts. Even Pukeko, *Porphyrio porphyrio*, have been seen wandering in the snow in these areas.

Southern Tokoeka   *Apteryx australis*

CHARACTERISTICS	Length — 40 cm. Voice of male sounds like a shrill *ki-wi*. Voice of female a hoarse shriek.
BIOLOGY	Food consists of worms, grubs, spiders and fallen fruit. One or two chicks hatch in spring after an incubation period of up to 80 days.
DISTRIBUTION	In mountains southeast of Haast.
HABITAT	A nocturnal species that lives in forest, scrub and tussocklands.
NOTES	A distinctive race of the brown kiwi that may deserve recognition as a full species. This population is threatened as it now numbers fewer than 300 birds.

Great Spotted Kiwi/Roa   *Apteryx haastii*

CHARACTERISTICS	Length — 50 cm. Voice a high-pitched, vibrating whistle.
BIOLOGY	Food similar to that of the Tokoeka. One or two chicks hatch in spring or early summer, after an incubation period of up to 85 days.
DISTRIBUTION	Northwest Nelson, northern Westland south to about the Buller River, the Paparoa Ranges and the Southern Alps from Arthur's Pass to the Hurunui River.
HABITAT	A nocturnal species living in forest, subalpine scrub, and occasionally in tussockland.
NOTES	The largest kiwi. Can often be heard calling at night. In recent years this kiwi has disappeared from a significant part of its range.

Southern Tokoeka — male.

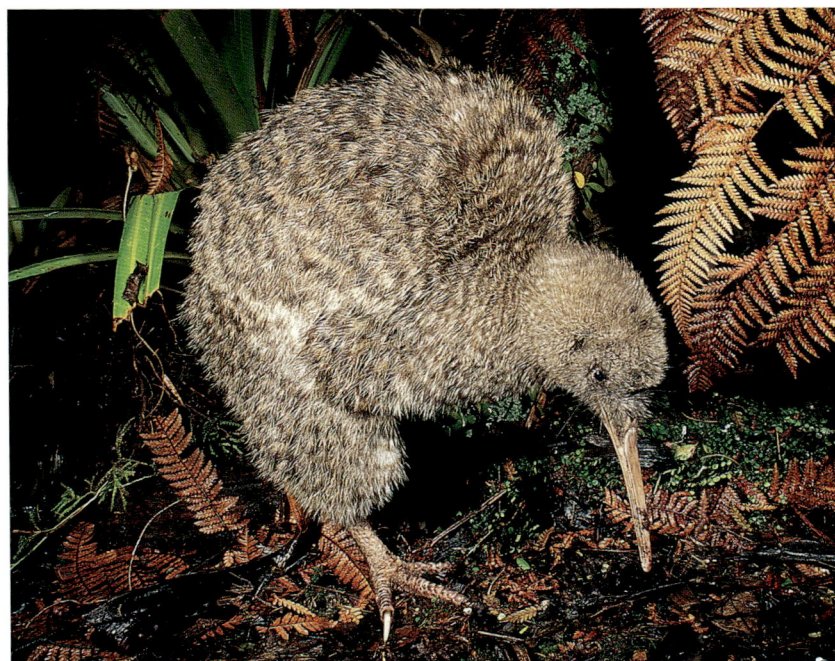

Great Spotted Kiwi — adult male.

Hutton's Shearwater/Titi *Puffinus huttoni*

CHARACTERISTICS	Length — 36 cm. Voice a loud cackling call.
BIOLOGY	Food consists of small fish and squid caught at sea. Breeds in December and January, raising a single chick.
DISTRIBUTION	Breeds at high altitudes in the Seaward Kaikoura Range. Ranges along the Kaikoura Coast, into Cook Strait and the Tasman Sea.
HABITAT	A seabird that nests high inland in burrows. Returns to its nest at night.
NOTES	Once more widespread, but numbers have declined due to predation by stoats. Often heard calling far overhead at night.

Canada Goose *Branta canadensis*

CHARACTERISTICS	Length — 80 cm. Voice in flight, or on the water, is a melodious double honk.
BIOLOGY	Food is generally grasses and seeds. One to ten chicks hatch from September to November after an incubation period of around 30 days.
DISTRIBUTION	In the North Island, at widely scattered locations south of Auckland. Throughout the high-country of the South Island. Many geese migrate to Lake Ellesmere and other coastal lakes in winter.
HABITAT	On tussocklands and near high-country lakes.
NOTES	Introduced from North America between 1876 and 1920. In non-breeding season can congregate in very large flocks and cause considerable damage to pastureland by grazing and fouling pasture.

Hutton's Shearwater pair in breeding colony.

Canada Goose — group of adults in riverbed.

Blue Duck/Whio *Hymenolaimus malacorhynchos*

CHARACTERISTICS | Length — 53 cm. Voice of male a melodious *whio*, of female a characteristic rasping *cr-ack*.

BIOLOGY | Food is insects and aquatic larvae taken by dabbling under and around rocks in water. Breeding takes place from August to September. Up to eight chicks are hatched.

DISTRIBUTION | In the North Island it is found on rivers of the Volcanic Plateau, Mount Egmont/Taranaki and at a few locations between the Urewera National Park and the Tararua Ranges. In the South Island it occurs in mountain streams from Nelson south to Fiordland, but is commoner on the western side of the Southern Alps.

HABITAT | Fast-flowing, turbulent streams and rivers.

NOTES | A threatened species whose range has declined dramatically since humans arrived in New Zealand.

Paradise Shelduck/Putangitangi *Tadorna variegata*

CHARACTERISTICS | Length — 63 cm. Voice is a duet performed in the air or from some high point. This is in two parts, with the male calling *zonk zonk*, followed almost immediately by the female calling *zeek zeek*.

BIOLOGY | Food consists of grasses and vegetation, also worms and insects. Nests from August to October, rearing up to ten chicks.

DISTRIBUTION | Widespread throughout New Zealand, including Stewart Island and the larger offshore islands. Very common in the eastern foothills of the Southern Alps. Often visits alpine tussocklands in summer.

HABITAT | On arable land, hill-country pasture and tussocklands. Often in the vicinity of farm ponds and tarns.

NOTES | Usually found in pairs or family groups. After breeding, sometimes flocks in large numbers onto cultivated lands to feed.

Blue Duck

Paradise Shelducks

Australasian Harrier/Kahu *Circus approximans*

CHARACTERISTICS	Length — 60 cm. Call is a shrill *kee-kee* during the breeding season but otherwise silent.
BIOLOGY	Food is birds, rodents and large insects. Frequently feeds on carrion. Two to three chicks are reared between late August and November.
DISTRIBUTION	Widespread throughout mainland New Zealand from sea level to alpine areas. Also found on the Chatham and Kermedec islands.
HABITAT	Open pastureland, tussockland and short scrub.
NOTES	One of the few native bird species that has benefited from human arrival in New Zealand, mainly because of the number of animals killed on roads by cars.

New Zealand Falcon/Karearea *Falco novaeseelandiae*

CHARACTERISTICS	Length — 45 cm. Call is a rapidly repeated *kek-kek-kek*.
BIOLOGY	Food is live large insects, rodents and small birds. Does not take carrion. Breeds from September to December, usually rearing two chicks.
DISTRIBUTION	At all altitudes from the Hunua Range south to the Auckland Islands. Nowhere common, and absent from much of Canterbury and Southland.
HABITAT	High-country pastureland and low scrub, sometimes in forest. Also over tussocklands. Perches high on trees, cliffs or rocky outcrops, watching for prey.
NOTES	An uncommon species. Occurs in three distinct forms — the Bush Falcon, the Eastern Falcon and the Southern Falcon.

RM

ABOVE:
Australasian Harrier —
adult.

RIGHT:
New Zealand Falcon

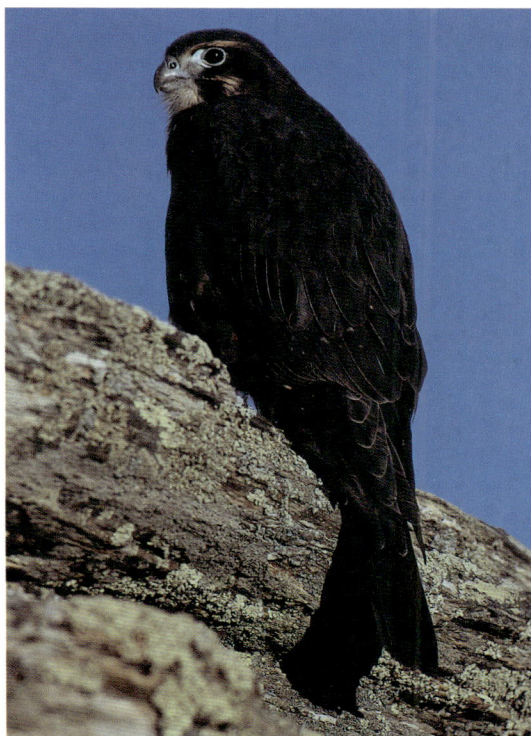

RM

Chukor *Alectoris chukar*

CHARACTERISTICS	Length — 33 cm. Voice is a fowl-like cackle or cluck.
BIOLOGY	Food is mainly berries and shoots, but also sometimes insects. Between ten and eighteen eggs are laid between September and February, but chick attrition is high.
DISTRIBUTION	The eastern side of the Southern Alps from Nelson south to Central Otago. Also, small populations near Tauranga and in Hawkes Bay. Occurs naturally from southeast Europe to China.
HABITAT	Open tussocklands and areas of scrub and rock with a preference for higher, drier areas.
NOTES	Chukor are descended from two distinct races which were introduced from India and Iran and are presumably hybrids.

Weka *Gallirallus australis*

CHARACTERISTICS	Length — 55 cm. Voice a drawn-out *week-ee-week*, but sometimes a shrill *coo-eet*.
BIOLOGY	Food consists of insects, worms, fledgling birds and rodents. Also scavenges. Nesting occurs throughout the year. Up to five young are reared.
DISTRIBUTION	In the North Island, Weka are now found mostly on the East Cape with a few introduced populations on islands such as Kawau and Pakatoa. In the South Island they are found mainly in Nelson, northern Westland and Fiordland. Also on Stewart Island and introduced to the Chatham Islands.
HABITAT	Areas of forest and scrub and in open country if cover is available.
NOTES	Usually remains in clearly defined territories. A wily bird which raids campsites given the opportunity. Most populations of weka are rapidly declining, probably due to predation and disease.

Chukor

Western Weka

Takahe *Porphyrio mantelli*

CHARACTERISTICS	Length — 63 cm. Voice a low *oomf*, and sometimes a weka-like *coo-et*. Often calls in duet.
BIOLOGY	Omnivorous feeder, feeding mostly on vegetation but sometimes taking insects. Eggs are laid in October and November and in optimum conditions two chicks are reared. They stay with their parents for their first year.
DISTRIBUTION	Once widespread throughout both the North and South islands. Now confined to the Murchison and Stuart ranges west of Lake Wakatipu, along with a few offshore islands where they have been introduced.
HABITAT	Tussocklands, swamp and bush margins.
NOTES	Although much effort has been given to the protection and management of the Takahe, the population is still in decline. Chicks are preyed upon by Stoats, Hawks and probably Weka.

South Island Pied Oystercatcher/Torea
Haematopus ostralegus

CHARACTERISTICS	Length — 46 cm. Voice a shrill *hu-eep* which carries for some distance.
BIOLOGY	Inland the bird feeds on invertebrates, including insects and worms. Two to three eggs are laid between August and November.
DISTRIBUTION	Along coastlines throughout New Zealand, but in the South Island often found far inland. Inland birds migrate to the coast in winter.
HABITAT	Riverbeds, rough ground and farmland. Sometimes in tussocklands.
NOTES	Forms large feeding and roosting flocks.

Takahe

South Island Pied Oystercatcher — adult.

Spur-winged Plover *Vanellus miles*

CHARACTERISTICS	Length — 35 cm. Voice is a shrill, grating, staccato *kekekekek*.
BIOLOGY	Food is invertebrates such as worms, grubs and insects. Breeds from September to December but earlier in the warmer, more northern regions. Raises three to four young.
DISTRIBUTION	Widespread throughout New Zealand except Fiordland. Also found in the Kermadec and Chatham islands. Ranges from coastal areas upwards into subalpine and sometimes alpine zones.
HABITAT	Riverbeds, wetland margins, rough ground, fallow ground and farmland. Sometimes in tussocklands.
NOTES	Mostly found in pairs but sometimes forms small flocks in autumn and winter. A relatively recent arrival in New Zealand, first reported as breeding at Invercargill Airport in 1932. Very aggressive when breeding, mobbing passing birds and mammals, particularly Harrier Hawks.

Black Stilt/Kaki *Himantopus novaezealandiae*

CHARACTERISTICS	Length — 38 cm. Voice is a puppy-like *yep-yep-yep*.
BIOLOGY	Food is almost entirely small invertebrates, both aquatic and taken along stream and river edges. One to four chicks are raised annually, between September and December.
DISTRIBUTION	In the Mackenzie country of South Canterbury. In winter some birds move to coastal areas in both the North and South islands.
HABITAT	Breeds along lake margins, riverbeds and around the shores of ponds. Feeds in shallower waters such as the riffles of braided rivers.
NOTES	An endangered species that hybridises with the closely related Pied Stilt, *Himantopus leucocephalus*. Also threatened by predators such as feral cats and mustelids.

ABOVE:
Spur-winged Plover —
adult.

RIGHT:
Black Stilt

Banded Dotterel/Pohowera *Charadrius bicinctus*

CHARACTERISTICS	Length — 18 cm. Voice is a shrill, staccato *pit-pit* or a ringing *chip-chip*.
BIOLOGY	Food inland is chiefly worms, grubs and insects. Breeds between August and January, raising up to three chicks.
DISTRIBUTION	Birds nest on beaches, rivermouths, shingle riverbeds and occasionally in pastures and tussockland. Some dotterels head north to coastal estuaries in winter, but birds from the South Island high-country migrate to Australia.
HABITAT	Mudflats, sandy beaches, shingle riverbeds, short pastures and high tussocklands.
NOTES	Our commonest dotterel.

New Zealand Dotterel/Tuturiwhatu
Charadrius obscurus

CHARACTERISTICS	Length — 28 cm. Voice is normally a single, loud *prip*.
BIOLOGY	Food inland is normally worms and insects. Breeds from August to December, raising one to three chicks.
DISTRIBUTION	In the North Island this dotterel breeds at rivermouths and on permanent sandbars. On Stewart Island it breeds high up on mountains.
HABITAT	Mudflats, sandy beaches, shingle riverbeds, short pastures and high tussocklands.
NOTES	An endangered and declining species because of predation and habitat disturbance.

Banded Dotterel

New Zealand Dotterel

Wrybill/Nguluparore *Anarhynchus frontalis*

CHARACTERISTICS	Length — 20 cm. Voice is a soft, chattering *wik-a-wik*.
BIOLOGY	Food consists of small invertebrates such as worms, grubs, insects and spiders. Breeds between September and December, raising two chicks.
DISTRIBUTION	Breeds on Canterbury and Otago braided riverbeds, often far inland. In winter, moves to North Island estuaries.
HABITAT	On shingle riverbeds and in marshy areas. In winter frequents mudflats.
NOTES	Has a unique bill that curves to the right.

Black-backed Gull/Karoro *Larus dominicanus*

CHARACTERISTICS	Length — 60 cm. Voice is a laughing *yo-yo-yo-yo-yo-yo*.
BIOLOGY	Feeds on invertebrates but is also an opportunistic scavenger. Rears two to three chicks between October and December.
DISTRIBUTION	Common throughout New Zealand in coastal areas. Often strays well inland. Breeds in coastal colonies but also sometimes high inland, around tarns and lakes. Also found in South America, Australia and South Africa.
HABITAT	Beaches, estuaries and harbours. Often visits ploughed paddocks.
NOTES	A bird that has increased considerably in numbers since human arrival in New Zealand due to its ability to scavenge.

Wrybill — adult male about to incubate eggs.

Black-backed Gull — adult.

Black-billed Gull/Tarapunga *Larus bulleri*

CHARACTERISTICS	Length — 37 cm. Voice a shrill, guttural *korr*.
BIOLOGY	Food is mostly invertebrates but will occasionally scavenge. Two chicks are reared between October and December.
DISTRIBUTION	Patchy coastal distribution in the North Island, but found in fair numbers around Lake Rotorua. In the South Island this is the only gull that permanently resides around high-country lakes. In winter many move to the coast.
HABITAT	Frequents lakesides, riverbeds and estuaries.
NOTES	More wary of humans than its relative the Red-billed Gull, *Larus novaehollandiae*.

Black-fronted Tern/Tarapiroe *Sterna albostriata*

CHARACTERISTICS	Length — 30 cm. Voice a sharp *tit-tit-tit* when in flight.
BIOLOGY	Food is smallish invertebrates such as insects, taken by hawking. Two to three chicks are reared between October and December.
DISTRIBUTION	Canterbury and Otago east of the Southern Alps. In winter it moves to the estuaries and harbours of both the North and South islands.
HABITAT	Riverbeds, ploughed fields and lake edges. Nests on shingle riverbeds and on lake margins in small colonies.
NOTES	New Zealand's only inland tern.

Black-billed Gull — adult.

Black-fronted Tern — adult.

Kakapo   *Strigops habroptilus*

CHARACTERISTICS	Length — 63 cm. Voice consists of booming noises made by the male in breeding season, otherwise mostly silent.
BIOLOGY	Food consists of seeds, shoots and rhizomes. Two to four eggs are laid at infrequent intervals, usually in 'mast' years.
DISTRIBUTION	Once widespread throughout New Zealand, often at high altitudes. Now, probably survives only on islands with managed populations.
HABITAT	Areas of forest, scrub and tussockland bordering on forest.
NOTES	Although much effort has been put into the protection and management of the kakapo, its population is in decline due to its low breeding rate and predation.

Kea   *Nestor notabilis*

CHARACTERISTICS	Length — 46 cm. Call a loud *keaaa*, in flight. Softer, conversational whistling notes when on the ground.
BIOLOGY	Food includes fruits, grubs and insects. Will also scavenge for animal fats. Two to three chicks are reared annually, between August and January.
DISTRIBUTION	Alpine and subalpine areas of the South Island. Descends to sea level in Fiordland. Often visits habitation.
HABITAT	Areas of forest, scrub and tussock to high altitudes. Nests in rock crevices or under boulders.
NOTES	A mischievous, inquisitive and intelligent bird. Has a reputation for causing mayhem.

RIGHT:
Kakapo — young male.

BELOW:
Kea — adult female.

RM

RM

Rock Wren/Matuhi *Xenicus gilviventris*

CHARACTERISTICS	Length — 9 cm. Voice is a three-note call with the first note emphasised.
BIOLOGY	Food is chiefly insects and spiders but also takes the fruit of alpine plants in season. Up to three chicks are reared each year between September and December.
DISTRIBUTION	Alpine and subalpine areas of the South Island from Nelson south to Fiordland.
HABITAT	Screes, fellfields and areas of scrub-covered rocks.
NOTES	A true alpine bird, in winter it shelters among snow-covered rocks.

Pipit/Pihoihoi *Anthus novaeseelandiae*

CHARACTERISTICS	Length — 18 cm. Voice short, sharp *pith* or *cheet*.
BIOLOGY	Food is seeds and small invertebrates. Three to four chicks are hatched annually between September and January.
DISTRIBUTION	Open country from sea level to alpine areas. Throughout the mainland as well as on the Chatham and Subantarctic islands. Overseas, found from Africa through to Asia and Australia.
HABITAT	Pastureland, sand dunes, fellfields and tussocklands.
NOTES	Similar to the introduced Skylark, *Alauda arvensis*, except the Pipit does not soar.

RM

ABOVE:
ABOVE:
Rock Wren

RIGHT:
New Zealand Pipit

IS

Chaffinch *Fringilla coelebs*

CHARACTERISTICS	Length — 15 cm. Voice a bright, cheerful *pink*.
BIOLOGY	Food is mainly seeds and berries plus occasional insects. Three to five eggs are laid between September and January.
DISTRIBUTION	Widespread in all areas throughout New Zealand, from sea level to alpine areas.
HABITAT	In hedgerows, copses, scrub and along the edges of forest. Sometimes in pine plantations.
NOTES	In winter flocks together with other finches, often feeding on areas of stubble.

Yellowhammer *Emberiza citrinella*

CHARACTERISTICS	Length — 16 cm. Voice a metallic *twink*.
BIOLOGY	Food is mostly grass seeds, occasionally supplemented by insects and spiders. Three to four eggs are laid each year in the spring.
DISTRIBUTION	Widespread throughout New Zealand at all altitudes. Also on the Chatham and Kermadec islands.
HABITAT	In hedgerows, copses, scrub and along the edges of forest.
NOTES	In winter flocks together with other finches, often feeding on areas of stubble.

Chaffinch — adult male in winter.

Yellowhammer — adult male.

Goldfinch *Carduelis carduelis*

CHARACTERISTICS	Length — 12 cm. Voice a pleasant twittering *tswitt-witt-witt*.
BIOLOGY	Food is mostly seeds. Three to six eggs are laid between August and November.
DISTRIBUTION	Open country throughout New Zealand at all altitudes. Also on the Chatham and the Auckland islands.
HABITAT	Plantations, farmland, secondary growth, gardens, hedgerows and scrublands.
NOTES	Forms mixed feeding flocks in winter. Probably more common in New Zealand than in its native home of Britain because of the widespread use of pesticides there.

Redpoll *Carduelis flammea*

CHARACTERISTICS	Length — 12 cm. Voice a drawn-out metallic *tich-ich-ich-ich*.
BIOLOGY	Food is seeds and soft vegetation. Four to six chicks are reared annually from September to January.
DISTRIBUTION	From coastal to alpine areas throughout mainland and also on some subantarctic islands. More common in rural than urban areas.
HABITAT	Scrub, tussocklands, sand dunes and areas of weed-infested pasture.
NOTES	New Zealand's smallest finch. Forms large flocks in non-breeding periods, sometimes with other finch species.

ABOVE:
Goldfinch

RIGHT:
Redpoll — adult
male.

Frogs and Lizards

With around 60 lizards so far described, there are more lizards in New Zealand, in proportion to land area, than any other place on earth, and yet more species are being discovered all the time.

Although Australia has eight times the number of lizards we have, it is almost 30 times larger than New Zealand. Great Britain, which has roughly the same land area, has only three species. In New Zealand, lizards are found at higher latitudes and higher altitudes than those of any other country.

Greywacke rocks are efficient storers of solar energy and consequently have a surface temperature up to four degrees higher than that of the ambient temperature. This is an important factor in allowing our lizards to survive at high altitudes as they utilise this warmth for basking.

Archey's Frog *Leiopelma archeyi*

CHARACTERISTICS	Length to 40 mm. Cryptic and nocturnal. Makes no sound.
BIOLOGY	Diet is chiefly small invertebrates, but also includes such things as fern spores. The tadpole does not have an aquatic phase.
DISTRIBUTION	Moehau and Coromandel ranges on the Coromandel Peninsula and the Whareorino Forest in the Herangi Range.
HABITAT	Damp forest, tussock clearings and subalpine scrub to about 1000 m. Found on mist-covered ridges under logs and stones.
NOTES	An endangered frog.

Whistling Frog *Litoria ewingii*

CHARACTERISTICS	Length to 70 mm. Nocturnal. Loud peeping call at night.
BIOLOGY	Diet consists of small invertebrates.
DISTRIBUTION	Widespread in the South and Stewart islands and in the Manawatu in the North Island. Small pockets of Whistling Frogs occur elsewhere, in places such as Wanganui in the North Island.
HABITAT	From coastal forest to subalpine tussock and scrub.
NOTES	Introduced to Westland from Tasmania in the 1870s.

Black-eyed Gecko *Hoplodactylus kahutarae*

CHARACTERISTICS	Length — 28 cm. Basks in sun on warm days.
BIOLOGY	Diet is chiefly invertebrates but probably also includes the fruits of some alpine plants.
DISTRIBUTION	Seaward and Inland Kaikoura Ranges.
HABITAT	Crevices, rockfaces and bluffs at altitudes 1300–2200 m.
NOTES	Can survive at higher altitudes and lower temperatures than any other New Zealand lizard.

RM

Archey's Frog — male with single froglet.

RM

Whistling Frog — adult.

RM

Black-eyed Gecko — female.

Remarkables Gecko   *Hoplodactylus* sp.

CHARACTERISTICS	Length to 10 cm. Cryptic but often basks by day.
BIOLOGY	Diet is presumably similar to that of the Black-eyed Gecko, *Hoplodactylus kahutarae*.
DISTRIBUTION	Restricted to the subalpine and alpine zones of the Remarkables, the Hector and Eyre mountains and the Slate Range mountains to the south and east of Lake Wakatipu.
HABITAT	Scree slopes and rocky outcrops in tussock and shrublands.
NOTES	So far not scientifically described.

Takitimu Gecko   *Hoplodactylus* sp.

CHARACTERISTICS	Length to 14 cm. Diurnal but cryptic.
BIOLOGY	Diet presumably the same as that of the Black-eyed Gecko.
DISTRIBUTION	Restricted to the subalpine and alpine zones of the Takitimu Ranges in western Southland.
HABITAT	Scree slopes and rocky outcrops in tussock and shrublands.
NOTES	So far not scientifically described.

Grand Skink   *Oligosoma grande*

CHARACTERISTICS	Length to 26 cm. Basks on sunny days on exposed tors and schist outcrops.
BIOLOGY	Invertebrates and fruits and berries in season. Gives birth to about a dozen young in late spring.
DISTRIBUTION	Widespread in Otago but nowhere common.
HABITAT	Shrublands and tussocklands with rocky outcrops and tors.
NOTES	An endangered species, whose range has contracted severely since human arrival due to land clearance and predation. A diurnal species that takes cover when it is disturbed.

Remarkables Gecko

Takitimu Gecko

Grand Skink — adult female.

Otago Skink *Oligosoma otagense*

CHARACTERISTICS	Length to 30 cm. Basks on sunny days on tors and schist outcrops.
BIOLOGY	Diet is mostly invertebrates but also fruit and berries in season. Breeding biology similar to that of the grand skink.
DISTRIBUTION	Widespread but uncommon in Otago.
HABITAT	Tors and rocky outcrops in tussocklands and dry shrublands.
NOTES	Another endangered species, whose range has contracted severely since human arrival due to land grazing and predation. A diurnal species that often basks in the sun, but which rapidly takes cover if disturbed.

Scree Skink *Oligosoma waimatense*

CHARACTERISTICS	Length to 30 cm. Basks on rocks on sunny days, but rapidly takes cover if disturbed.
BIOLOGY	Diet is mainly insects, spiders and other small invertebrates. Breeding biology is not known.
DISTRIBUTION	Inland areas on the eastern side of the Main Divide from Marlborough south through Canterbury to northern Otago.
HABITAT	Scree slopes, dry streambeds and rocky outcrops in drier tussocklands to about 1500 m.
NOTES	Widespread but uncommon.

Otago Skink — adult female.

Scree Skink

Fish

Our native fishes have been adversely affected by the introduction of exotic species and although only one species, the Grayling or Upokororo, *Proctotroctes oxyrhynchus*, is known to have become extinct, there were probably discrete and cryptic galaxiid species that disappeared before being classified.

Around 40 species of galaxiid or Galaxidae fishes are found throughout the southern hemisphere, and many of these are closely related to species living here. In fact, one of our commonest species, the Inanga, *Galaxias maculatus*, is widespread throughout the southern hemisphere from the Falkland Islands, through Tierra del Fuego, the Subantarctic Islands, New Zealand, southern Australia and Tasmania to South Africa. This is taken as yet another indication of our ancient Gondwanaland connections.

Other fish that often occur in alpine streams and lakes are the bullies belonging to the Eleotridae, a family that is widespread and common in fresh and saltwater in the tropical Pacific and South-East Asia, but which reaches its southern limit in New Zealand, on Stewart Island.

The Torrentfish, another fish often found in alpine areas, is the sole freshwater member of the Pinguipedidae, a family to which our familiar Blue Cod belongs.

Koaro   *Galaxias brevipinnis*

CHARACTERISTICS	Cryptic and usually nocturnal. Length usually 160–180 mm, but a fish of 271 mm has been recorded.
BIOLOGY	Food generally consists of simuliid and chironomid larvae, along with the larvae of mayfly, stonefly and caddisfly, but other small insects are taken if opportunity allows. Several thousand eggs are laid in autumn and early winter in marginal gravel and vegetation. The larvae, on hatching, go to sea for about 20 weeks, returning as part of the spring whitebait 'run'.
DISTRIBUTION	Widespread throughout New Zealand from sea level to alpine areas. Sometimes found in landlocked alpine lakes.
HABITAT	In fast-flowing, bouldery streams usually flowing through forest. Lurks under boulders, logs and overhanging banks.
NOTES	The young of the Koaro are among galaxiid species which make up our whitebait catch.

Alpine Galaxias   *Galaxias paucispondylus*

CHARACTERISTICS	Cryptic and generally nocturnal. The average alpine galaxias is about 80–85 mm, but individuals of up to 112 mm have been found.
BIOLOGY	Food requirements have not been studied but are presumably similar to those of the Koaro, *Galaxias brevipinnis*.
DISTRIBUTION	At a few scattered localities along the eastern side of the Southern Alps, as well as in the catchments of the Maruia, Buller and Wairau rivers on the West Coast and the Oreti River in Southland.
HABITAT	The rapids of fast-flowing, often snow-fed, gravelly/bouldery mountain streams.
NOTES	The female lays 100–300 eggs in the spring but, unlike many other native galaxiids, the alpine galaxias does not have an ocean-going larval stage.

RM

Koaro — adult.

RMM

Alpine Galaxias

Upland Bully/Pako *Gobiomorphus breviceps*

CHARACTERISTICS	Diurnal. This bully averages 85–90 mm but can on occasion exceed 100 mm.
BIOLOGY	Food consists chiefly of the larvae of aquatic insects but crustaceans and snails are also taken. The female spawns every few weeks during the summer with the eggs hatching after several weeks.
DISTRIBUTION	The Wanganui, Manawatu and Ruamahanga rivers of the southern North Island and in the South and Stewart islands from sea level up into the alpine zone.
HABITAT	A variety of habitats from gravelly rivers and streams to weedy lakes.
NOTES	The most widespread of the native bullies.

Torrentfish/Papanoko *Cheimarrichthys fosteri*

CHARACTERISTICS	Diurnal. The average size of the Torrentfish is 100–125 mm but fish exceeding 200 mm have been reported.
BIOLOGY	Generally grazes on algae-covered rocks. Occasionally takes aquatic insect larvae. The female lays several thousand eggs, probably in late summer, and the larval stage of the Torrentfish is spent at sea.
DISTRIBUTION	Throughout the North and South islands from sea level up to moderately high altitudes, as far inland as it can migrate from the coast.
HABITAT	Fast-flowing rivers and larger streams usually in, or near, rapids.
NOTES	The larval stage of the Torrentfish is spent at sea.

Upland Bully — male.

Torrentfish — adult.

Insects

On spring and summer days, insects are a conspicuous feature of our upland areas where they are present in an amazing array and diversity. Alpine insects are often closely related to lowland species, but are usually larger, darker and hairier so as to cope with the harsh conditions. They are all able to survive the winters in larval or pupal stages.

Day-flying or diurnal moths are often encountered in our mountains with over 100 species of the colourful geometrid moths being found here.

Cicadas, which belong to an endemic genus the *Maoricicada*, are obvious elements in the alpine zone because of the strident mating songs of the males. Other common insects in this zone are the members of the Orthoptera order such as grasshoppers, weta and crickets. Weta are often thought of as being an endemic group but are actually also found in a number of other countries. However, nowhere else do they achieve the size and often spectacular colouration of those species found in New Zealand.

Spiny-gilled Mayfly/Piri Wai *Coloburiscus humeralis*

CHARACTERISTICS	Length — 10 mm. Wings are held 'sail-like' over the body when the insect is resting.
BIOLOGY	Has an aquatic larval stage.
DISTRIBUTION	Montane to alpine areas from Northland south to Stewart Island.
HABITAT	The undersides of stones and boulders in fast-moving, clear, stony streams.
NOTES	Adult mayflies have a short life of only a couple of days. During this period they swarm together at dawn and dusk to mate. A member of the Ephemeroptera order.

Mountain Giant Dragonfly/Kapokapowai *Uropetala chiltoni*

CHARACTERISTICS	Length to 86 mm. When resting the wings are held out horizontally.
BIOLOGY	The larval stage is aquatic and may last up to six years. Food consists of small insects taken in flight.
DISTRIBUTION	From montane to alpine zones in the Seaward Kaikouras and along both sides of the Main Divide of the South Island. There is only one record in the North Island, from near Otaki.
HABITAT	Around tarns, swamps and along bog margins, as well as along the quieter upland stretches of streams and rivers.
NOTES	Adults emerge from early summer through to autumn. A member of the Odonata order.

Alpine Green Stoner *Acroperla christinae*

CHARACTERISTICS	Length — 13 mm. Almost always found near fresh water.
BIOLOGY	Larvae are aquatic and carnivorous, feeding on the nymphs of other aquatic insects.
DISTRIBUTION	Alpine areas of South Westland and west of the Main Divide in western Otago.
HABITAT	Swift alpine streams flowing through snowgrass and shrubland.
NOTES	A member of the Plecoptera order.

Spiny-gilled Mayfly

Mountain Giant Dragonfly

Alpine Green Stoner

Giant Stoner *Holcoperla magna*

CHARACTERISTICS	Length to 24 mm. A flightless species.
BIOLOGY	The larvae are aquatic and carnivorous, feeding on the nymphs of other aquatic insect species.
DISTRIBUTION	High-alpine areas in the mountains of western Otago.
HABITAT	Headwaters of high-alpine streams. Larvae are found under rocks while the large, flightless adults are found in the dry spaces higher up under rocks in these same streams.
NOTES	There are two species of giant stoners in the genus *Holcoperla*, with *H. angularis* replacing this species in the Takitimu Mountains and in the mountains of Fiordland.

Alpine Running Stoner *Zelandobius atratus*

CHARACTERISTICS	Length to 9 mm. A flightless species, often active by day, crawling over streamside vegetation. Very active, sheltering under streamside rocks and vegetation.
BIOLOGY	Aquatic larval stage with larvae feeding on nymphs of other aquatic insects.
DISTRIBUTION	High-alpine areas in the mountains of Otago.
HABITAT	Headwaters of streams.
NOTES	Both sexes are short-winged and flightless. Sometimes active by day, crawling on streamside vegetation. A species-rich genus of 27 animals, found in most alpine areas.

Alpine Cockroach/Papata *Celatoblatta quinquemaculata*

CHARACTERISTICS	Length to 12 mm. A nocturnal and fast-moving species.
BIOLOGY	Food is unknown, but probably the detritus found among leaf litter.
DISTRIBUTION	High-alpine areas in the mountains of Otago.
HABITAT	Large colonies are found in and among rocks in fellfields.
NOTES	Similar, related species are found in other alpine areas such as those in Fiordland, Southland and Canterbury. A member of the Blattodea order.

BP

ABOVE:
Giant Stoner

RIGHT:
Alpine Running Stoner

BELOW:
Alpine Cockroach

BP

BS

Mountain Tree Weta *Hemideina maori*

CHARACTERISTICS	Length to 55 mm. Nocturnal.
BIOLOGY	Generally carnivorous but will take some vegetable matter. Although eggs can be laid throughout the year, April and May are the preferred months, with the larvae hatching the following summer.
DISTRIBUTION	Low- and high-alpine zones from Mount Cook south to northern Otago.
HABITAT	Rocky areas and fellfields.
NOTES	Several colour forms of this weta have been recorded. A member of the Orthoptera order.

Scree Weta *Deinacrida connectens*

CHARACTERISTICS	Length to 65 mm. Nocturnal.
BIOLOGY	Similar to that of the Mountain Tree Weta, *Hemideina maori*.
DISTRIBUTION	Alpine areas of the South Island from Mount Arthur in the Kahurangi National Park in Nelson, south to the Takitimu Mountains in Southland.
HABITAT	Rocky areas and scree slopes.
NOTES	This weta is replaced in the mountains to the west of the Main Divide by the Alpine Weta, *Deinacrida pluvialis*.

Bluff Weta *Deinacrida elegans*

CHARACTERISTICS	Length to 50 mm. Nocturnal. Skilled rock climber.
BIOLOGY	Generally little known.
DISTRIBUTION	Montane to low-alpine areas in the Seaward Kaikouras, as well as on Mount Somers in mid-Canterbury.
HABITAT	In the narrow crevices and fissures of rocky bluffs.
NOTES	One of the most spectacular of our weta species.

Mountain Tree
Weta — male.

Scree Weta —
male.

Bluff Weta —
male.

Snowbank Grasshopper/Kowhitiwhiti
Alpinacris tumidicauda

CHARACTERISTICS	Length to 20 mm. Diurnal.
BIOLOGY	Herbivorous, feeding mostly on grasses. Eggs are deposited underground with the length of the larval stage dependent on the availability of food.
DISTRIBUTION	Alpine areas in the drier eastern mountains of Central Otago, northern Southland and eastern Fiordland.
HABITAT	Usually near snowbanks where they feed on herbs and foliage.
NOTES	Distinguished by its orange hind-parts. A member of the Orthoptera order.

Mountain Grasshopper/Kowhitiwhiti
Sigaus campestris

CHARACTERISTICS	Length to 16 mm. Diurnal
BIOLOGY	Little known but presumably similar to that of the Snowbank Grasshopper, *Alpinacris tumidicauda*.
DISTRIBUTION	Montane to alpine zones in the mountains and valleys of the South Island, from Canterbury south to northern Southland.
HABITAT	In grasslands and near snowbanks.
NOTES	Often locally abundant. An important defoliator of vegetation.

Scree Grasshopper   *Brachaspis nivalis*

CHARACTERISTICS	Length to 18 mm. Diurnal.
BIOLOGY	Not known but presumably similar to that of the Snowbank Grasshopper, *Alpinacris tumidicauda*.
DISTRIBUTION	Widespread throughout the mountains of the central South Island from the Black Birch Mountains of Marlborough and the Kaikoura Ranges south to around Lake Ohau.
HABITAT	Rocky areas including screes of the alpine areas.
NOTES	The cryptically coloured adults can be locally common. There are three described species in the genus: *Brachaspis nivalis* and *B. robustus* which occur from the Mackenzie country south to Central Otago, and *B. collinus* from upland areas in the North Island.

Snowbank
Grasshopper

Mountain
Grasshopper

Scree
Grasshopper

Nival Cicada/Kihikihi *Maoricicada nigra*

CHARACTERISTICS	Length to 12 mm. Diurnal. In spring, males sing loudly from vantage points with a characteristic chattering song.
BIOLOGY	Herbivorous. The nymphs spend several years underground, feeding on the roots of plants.
DISTRIBUTION	Subalpine and alpine zones of the South Island's Main Divide from the Victoria Range in Nelson south to Fiordland and east to Central Otago.
HABITAT	Rocky areas, herbfields and fellfields, from which the males sing.
NOTES	A member of the Hemiptera order.

Scree Tiger Beetle/Papapa *Neocicindela hamiltoni*

CHARACTERISTICS	Length to 20 mm. Diurnal.
BIOLOGY	Beetles run quickly across screes searching out prey. Larvae construct tunnels and wait by the entrance for passing insects.
DISTRIBUTION	Widespread from coastal to alpine areas of the South Island from Marlborough south to Central Otago.
HABITAT	In sandy pockets close to screes and on more open rocky areas.
NOTES	A member of the Coleoptera order.

Bumbling Chafer/Tataka *Scythrodes squalidus*

CHARACTERISTICS	Length to 18 mm. Diurnal.
BIOLOGY	The larval stage is spent underground. Larvae feed on the roots of plants.
DISTRIBUTION	Low- and high-alpine zones in the mountains of Otago.
HABITAT	Herbfields and grasslands, often at the upper limits of vegetation.
NOTES	Adults are active by day, bumbling over vegetation and sometimes climbing tall herbs to feed. A member of the Coleoptera order.

Nival Cicada
— male.

RM

Scree Tiger
Beetle — adult.

RM

Bumbling
Chafer

BP

Giant Black and White Weevil *Lyperobius hudsoni*

CHARACTERISTICS	Length to 22 mm. Diurnal.
BIOLOGY	Both adults and larvae feed on speargrasses, namely members of the carrot family, the Umbelliferae. Eggs are laid in the spring and the larval stage is believed to last about a year.
DISTRIBUTION	Alpine zones of many eastern South Island mountains south to South Canterbury, as well as on coastal cliffs near Wellington.
HABITAT	Alpine grassland in association with the speargrass species on which both adults and larvae feed.
NOTES	The Wellington population is threatened with extinction due to habitat loss. A legally protected species. A member of the Coleoptera order.

Giant Alpine Weevil *Lyperobius spedeni*

CHARACTERISTICS	Length to 24 mm. Slow-moving and diurnal.
BIOLOGY	Adults and larvae feed on the small speargrass *Aciphylla lecomtei*. The larval stage probably lasts about a year.
DISTRIBUTION	Alpine areas in western and Central Otago as well as in western Southland and South Westland.
HABITAT	Grasslands and shrublands where speargrass species predominate.
NOTES	There are sixteen species in the genus.

Scorpionfly *Nannochorista philpotti*

CHARACTERISTICS	Length to 10 mm. Diurnal.
BIOLOGY	Eggs are laid in the spring and the larval stage is believed to last about a year.
DISTRIBUTION	From lowland to alpine areas in the South Island and on Stewart Island.
HABITAT	Slow-moving, muddy-bottomed streams where the scorpionfly larvae feed on midge larvae.
NOTES	New Zealand's only scorpionfly and one of the only scorpionfly species in the world to have aquatic larvae. A member of the Mecoptera order.

Giant Black
and White
Weevil

BP

Giant Alpine
Weevil

RM

Scorpionfly

BP

Bat-winged Fly *Exsul singularis*

CHARACTERISTICS	Wingspan to 28 mm. Diurnal, sunning itself on rock faces. Flight clumsy and close to ground.
BIOLOGY	Adults are insectivorous, feeding mostly on flies and butterflies. Larval stage not known.
DISTRIBUTION	Alpine areas west of the Main Divide from the Paparoa Range in Central Westland, south to Fiordland.
HABITAT	Swiftly flowing streams where adults predate emerging aquatic insects.
NOTES	An endemic genus. A member of the Diptera order.

Picture Fly *Tephritis cassiniae*

CHARACTERISTICS	Wingspan to 8 mm. Diurnal.
BIOLOGY	Larvae feed within the swellings on the stems of the cottonwood shrubs, *Ozothamnus* spp., but are otherwise little known.
DISTRIBUTION	Not known.
HABITAT	Alpine shrubland.

Alpine Black Caddis *Periwinkla childii*

CHARACTERISTICS	Length to 8 mm. Active at dusk and after nightfall.
BIOLOGY	The larvae construct cases from a silk-like material and live within these under water but the exact length of this larval stage is not known.
DISTRIBUTION	Alpine areas in the mountains of Otago.
HABITAT	Moderately fast-flowing, stony or rocky streams.
NOTES	A day-flying species of snow-melt areas. A member of the Trichoptera order.

Bat-winged Fly — adult with Southern Blue Butterfly.

Picture Fly

Alpine Black Caddis

Alpine Caddis *Philoreithrus lacustris*

CHARACTERISTICS	Length — 18 mm. Active at dusk and after nightfall.
BIOLOGY	Presumed to be similar to that of the Alpine Black Caddis, *Periwinkla childi*.
DISTRIBUTION	Alpine areas in the mountains of Otago.
HABITAT	Moderately fast-flowing, stony or rocky streams.
NOTES	A day-flying species that emerges as the snow melts.

Tussock Butterfly/Mokarakare
Argyrophenga antipodum

CHARACTERISTICS	Wingspan to 46 mm. Diurnal. Common from spring to autumn.
BIOLOGY	The larvae are cryptic and feed on snowgrasses during late spring and summer.
DISTRIBUTION	Widespread in coastal to alpine zones in the South Island, but absent from western Fiordland.
HABITAT	Associated with areas of snowgrass.
NOTES	There are three species in the genus, with *Argyrophenga janitae* being common in the eastern South Island from Nelson south to northern Southland, and the rare *A. harrisi* occurring from north-west Nelson to Lewis Pass. A member of the Lepidoptera order.

Black Mountain Ringlet *Percnodaimon merula*

CHARACTERISTICS	Wingspan to 40 mm. Diurnal. Rapid flyer.
BIOLOGY	The larvae feed on various small grasses growing in rocky habitats over the summer months.
DISTRIBUTION	Generally widespread in the mountains of the South Island, apart from much of Central Otago.
HABITAT	Low- and high-alpine rocky areas and also on scree slopes.

Alpine Caddis

Tussock
Butterfly

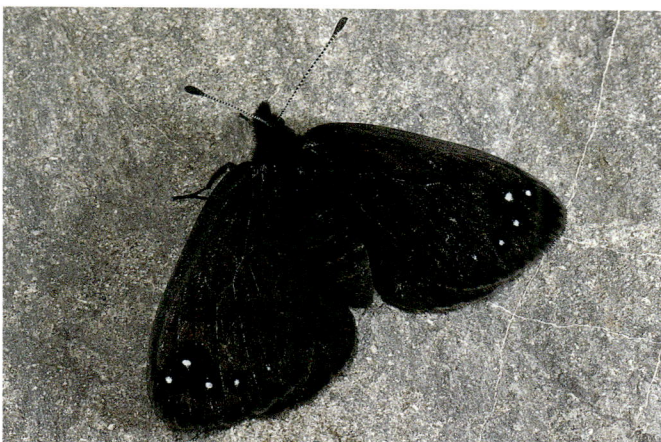

Black
Mountain
Ringlet

Black and Yellow Kelleria Looper   *Notoreas mechanitis*

CHARACTERISTICS	Wingspan to 30 mm. Diurnal.
BIOLOGY	The larvae feed on various species of kelleria cushions, *Kelleria* spp.
DISTRIBUTION	Western areas in the South Island mountains from Nelson to Fiordland.
HABITAT	Subalpine to high-alpine grassland and herbfields.

Giant Speargrass Moth   *Graphania nullifera*

CHARACTERISTICS	Wingspan to 62 mm. Nocturnal species that is attracted to lights.
BIOLOGY	Larvae feed within the taproots of various speargrass *Aciphylla* spp., sometimes killing the host plants in the process.
DISTRIBUTION	More coastal in the North Island, but widespread from lowland to high-alpine zones in the South Island.
HABITAT	Rocky areas in grasslands.

Black and Yellow
Kelleria Looper

Giant Speargrass
Moth

BP

BP

Alpine Ghost Moth *Aoroia flavida*

CHARACTERISTICS	Wingspan to 50 mm. A nocturnal species that is attracted to light.
BIOLOGY	The larvae spend their days underground, emerging at night to feed on litter and grasses. The females are flightless.
DISTRIBUTION	Known only from the Ajax Plateau, the Catlins and the Umbrella Mountains of northern Southland.
HABITAT	Alpine wetlands.
NOTES	One of thirteen species of *Aoroia* that are found from the central North Island through to Stewart Island.

Alpine Tiger Moth *Metacrias erichrysa*

CHARACTERISTICS	Wingspan to 36 mm. Diurnal. Flight fast and direct.
BIOLOGY	Larvae are very hairy, orange-brown in colour and feed on low-growing grasses and herbs. Female is small, dirty-white in colour and flightless.
DISTRIBUTION	Higher areas in the Tararua and Ruahine Ranges in the North Island and alpine areas of the western South Island from Nelson south to Fiordland.
HABITAT	Alpine grasslands and herbfields.
NOTES	Three species of the genus are found in New Zealand, and another three species occur in Tasmania and southeastern Australia.

Alpine Ghost Moth

Alpine Tiger Moth — male.

.

Spiders

L ike the alpine insects, our spiders found at these higher altitudes are little known and many mountain-dwelling species have yet to be closely examined and classified.

The main groups of spiders to be found above the timberline are the jumping spiders of the Salticidae family, the wolf spiders which belong to the Lycosidae, the stealthy or vagabond spiders which are members of the Gnaphosidae, and a number of the crab spiders which are classified as Thomisidae. Apart from the Katipo, *Latrodectus katipo*, Maori tended to give all spiders the collective name Pungaiwerewere.

Spiders use a variety of methods to capture their prey. Some capture flying insects by the use of webs, some lurk in burrows waiting for their lunch to wander by and others actively hunt down their prey. Most alpine spiders belong to this last group, detecting their victims either by sight or by sensing the vibrations they emit.

During the colder weather they retreat to burrows and crevices to wait for more clement times to return. Species like the various jumping spiders weave dense silken cocoons in which they shelter from the winter cold.

White-bearded Mountain Jumping Spider
Salticidae sp.

CHARACTERISTICS	Length 4–6 mm. Diurnal. Prominent, square, headlight-like eyes.
BIOLOGY	Females build dense webs under rocks in late summer. They remain guarding the eggs, with the spiderlings emerging in autumn.
DISTRIBUTION	Subalpine and alpine zones of Otago and Southland.
HABITAT	Among scree and in rocky areas.
NOTES	Able to survive under snow for 4–5 months by sheltering in the dense silk cocoons it constructs under rocks.

Black Mountain Jumping Spider   *Salticidae* sp.

CHARACTERISTICS	Length 4–6 mm. Diurnal. Distinctive, large, forward-facing eyes.
BIOLOGY	Females build dense webs under rocks in late summer. They remain guarding the eggs, with the spiderlings emerging in autumn.
DISTRIBUTION	Subalpine and alpine zones of Otago and Southland and possibly South Canterbury.
HABITAT	In rocky areas, under boulders and among scree.
NOTES	Able to survive under snow for 4–5 months, like the White-bearded Mountain Jumping Spider, by sheltering in the dense silk cocoons it constructs under rocks.

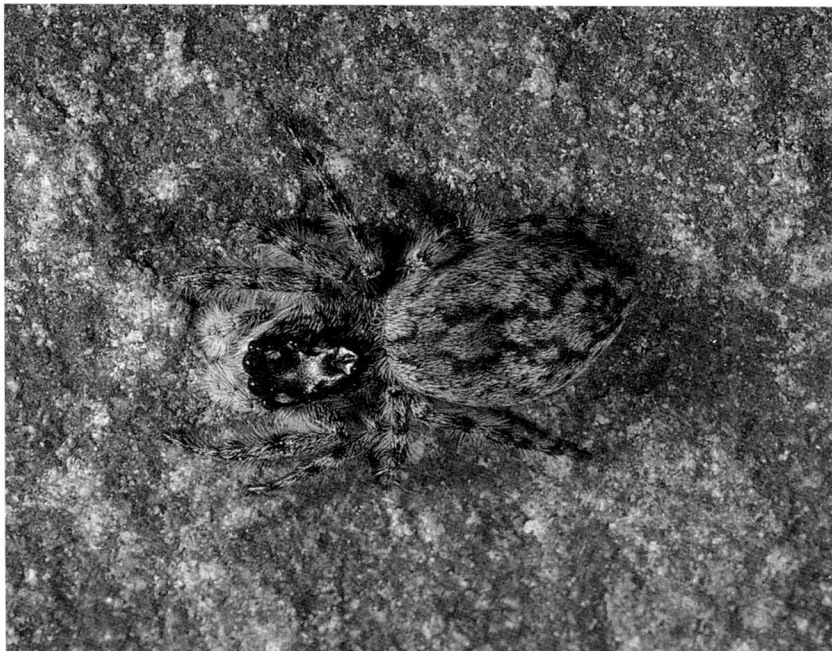

White-bearded Mountain Jumping Spider

Black Mountain Jumping Spider

Scree Prowling Spider   *Miturga* sp.

CHARACTERISTICS | Length 20–25 mm. Nocturnal. Male more slender and with longer legs than female.

BIOLOGY | The female prepares a silk-lined chamber under a rock or log and lays a single, large egg sac which she guards until the spiderlings hatch and disperse.

DISTRIBUTION | High-alpine areas in the mountains of Central Otago.

HABITAT | Rocky areas often in, or adjoining, *Raoulia* cushion plants. The small invertebrates on which this spider preys live here.

Green Crab Spider   *Diaea ambara*

CHARACTERISTICS | Length 2–4 mm. Diurnal. Forelegs held up crab-like in front of body.

BIOLOGY | The male 'ties down' the female while mating to prevent being eaten.

DISTRIBUTION | Montane and subalpine zones of Otago and Southland.

HABITAT | A free-living spider which lurks on bark and among the foliage of shrubs and trees in wait for its prey.

Scree Prowling Spider

Green Crab Spider

Molluscs

New Zealand has a great variety of snails, ranging in size from tiny species that could easily fit through the eye of a needle, to giants that would quite comfortably cover a small saucer.

Many of the large *Paraphanta* and *Powelliphanta* species are spectacularly coloured with striking markings. Some have only limited distributions, but our alpine species tend to range over large areas.

Although most snails are vegetarian, the large species found in the alpine areas are carnivorous, preying on giant earthworms which they seek out in the leaf litter.

There are about 25 species of leaf-veined slugs, many of which are very attractive with indentations that look like leaf patterns, and with flecks on their upper surfaces. During drier periods they hole up in damp, shady places such as rock overhangs and crevices.

All leaf-veined slugs are vegetarian, using their radulae to scrape algae off rocks and to dine on fungi.

Superb Giant Land Snail/Pupu Rangi
Powelliphanta superba

CHARACTERISTICS	Length 70–90 mm. Usually nocturnal but sometimes active after rain.
BIOLOGY	A vermivore which means that it feeds only on worms. Lays up to ten eggs each spring, which hatch six months later.
DISTRIBUTION	Subalpine to alpine areas of northwest Nelson, in and bordering on the Kahurangi National Park.
HABITAT	Widely but somewhat spasmodically distributed in beech forest, shrublands and amongst snowgrasses, *Chionochloa* spp.
NOTES	Predated by weka and introduced rats.

Leaf-veined Slug/Ngata *Pseudodoneitea papillata*

CHARACTERISTICS	Length to 30 mm. Nocturnal but sometimes active after rain. The upper surface of the animal carries a distinctive leaf-like pattern.
BIOLOGY	Feeds on fungi and probably rotting wood. Breeding biology not known.
DISTRIBUTION	From lowland to low-alpine zones in the North, South and Stewart islands, as well as on some offshore islands.
HABITAT	Often found in, or near, flax bushes, *Phormium* spp.

Superb Giant Land Snail

Leaf-veined Slug

Leaf-veined Slug/Ngata *Reflectopalium dalli*

CHARACTERISTICS	Length to 40 mm. Nocturnal. The upper surface of the animal carries a distinctive leaf-like pattern.
BIOLOGY	Feeds on fungi and probably rotting wood.
DISTRIBUTION	From lowland to subalpine areas from Pelorous Sound and adjacent areas of the South Island.
HABITAT	A nocturnal species, which is believed to feed on fungus and leaf litter.

Carnivorous Flatworm *Arthurdenyus testacea*

CHARACTERISTICS	Length to 400 mm. Chiefly nocturnal but sometimes emerges after rain.
BIOLOGY	Lives almost exclusively on earthworms.
DISTRIBUTION	Montane and subalpine areas from the Tararua Ranges southwards.
HABITAT	Lives in leaf litter and under logs.

Leaf-veined Slug — adult.

Carnivorous Flatworm

Shrubs

S hrubs can at times form dense thickets with interlacing branches that are the bane of the tramper trying to reach the alpine grasslands and herbfields higher up.

Above these tiresome thickets and out in the more exposed areas, shrubs have a tendency to grow low to the ground as prostrate or procumbent plants, or to hug the ground as mats or cushions to lessen wind resistance and consequently are much more readily negotiated.

Besides growing low to the ground, some other obvious adaptations shrubs have made to cope with the hostile climate are leaves that are either thick and fleshy or tough and leathery. Their leaves also often have a dense covering of hairs on one or both surfaces that serve to reduce transpiration.

As with the herbs, there is a high rate of endemism among the native shrubs. Prominent among these are the Hebe and Parahebe species, which are almost exclusively New Zealand groups. Some species in these groups also occur in a form known as a 'whipcord', that is more resistant to cold and dehydration.

Shrub Groundsel *Brachyglottis pentacopa*

CHARACTERISTICS	Height to 1 m.
BIOLOGY	Flowers from January to March.
DISTRIBUTION	Montane and alpine areas of the South Island from Marlborough south to Fiordland.
HABITAT	Bush edges, scrub and shrublands.
NOTES	A member of the daisy family, the Asteraceae.

Heather *Calluna vulgaris*

CHARACTERISTICS	Height to 50 cm.
BIOLOGY	Flowers from November to March.
DISTRIBUTION	Montane and subalpine areas of the Volcanic Plateau of the North Island, extending as far north as Rotorua.
HABITAT	Bush edges, scrub and shrublands preferring more open and less shaded sites.
NOTES	An introduced species that is now considered to be a pest as it smothers other vegetation. Also sometimes called Ling. A member of the northern heath family, the Ericaceae.

Mountain Cabbage Tree/Toii *Cordyline indivisa*

CHARACTERISTICS	Height to 4 m.
BIOLOGY	Flowers November and December.
DISTRIBUTION	Montane and subalpine areas of the North Island and Westland at altitudes 500–1100 m.
HABITAT	In scrub and along forest edges.
NOTES	The lowland Cabbage Tree, *Cordyline banksii* is similar but has narrower leaves. A member of the agave family, the Agavaceae.

Shrub
Groundsel

CM

Heather

CM

Mountain
Cabbage Tree

RM

Matagouri/Tumatakuru *Discaria toumatou*

CHARACTERISTICS	Height to 5 m.
BIOLOGY	Flowers in November and December. Fruits from January to March.
DISTRIBUTION	From the Wairarapa in the North Island south to Southland in montane and subalpine areas.
HABITAT	Stream and riverbeds, tussock lands, stony and rocky areas, occasionally in shrublands.
NOTES	Sometimes also known as the wild Irishman. Forms dense thickets that can be difficult to traverse. A member of the buckthorn family, the Rhamnaceae.

Large Grass Tree/Inaka *Dracophyllum traversii*

CHARACTERISTICS	Height to 13 m.
BIOLOGY	Flowers in January and February.
DISTRIBUTION	Montane and subalpine areas of the northern South Island from Nelson south to about Arthur's Pass.
HABITAT	Edges and clearings of subalpine forest and subalpine scrub.
NOTES	This is the largest of the native grass trees. In the North Island it is replaced by *Dracophyllum latifolium* which is very similar. A member of the southern heath family, the Epacridaceae.

Bush Snowberry/Koropuka *Gaultheria antipoda*

CHARACTERISTICS	Height 30 cm to 2 m.
BIOLOGY	Flowers from November to January. Fruits from January to April.
DISTRIBUTION	From coastal to subalpine zones, but more common at lower altitudes in the South Island and on Stewart Island.
HABITAT	In tussocklands, fellfields, rocky areas and in open scrub.
NOTES	Hybridises readily with Prostrate Snowberry, *Pernettya macrostigma*. A member of the northern heath family, the Ericaceae. Berries are edible.

Matagouri

CM

Large Grass Tree

RM

Bush Snowberry

JB

Large-flowered Hebe *Hebe macrantha*

CHARACTERISTICS	Height to 60 cm.
BIOLOGY	Flowers from December to February.
DISTRIBUTION	South Island mountains from subalpine to low-alpine zones at altitudes 1000–1500 m.
HABITAT	Steep dry rocky and stony places and shrublands. Only occasionally in tussocklands.
NOTES	A member of the figwort family, the Scrophulariaceae.

Ruapehu Hebe *Hebe venustula*

CHARACTERISTICS	Height to 1.5 m.
BIOLOGY	Flowers from December to February.
DISTRIBUTION	Eastern and central sections of the North Island from Mount Hikurangi to Mount Ruapehu in subalpine and alpine areas.
HABITAT	In scrublands, along forest margins and in alpine herbfields.

Pygmy Pine/Rimu *Lepidothamnus laxifolius*

CHARACTERISTICS	Forms mats up to 60 cm across.
BIOLOGY	Male strobili appear during November and December. Fruits in March and April.
DISTRIBUTION	From Tongariro National Park in the central North Island southwards to Stewart Island in montane and subalpine areas.
HABITAT	Scrub and fellfields where it forms dense turfs and mats sprawling over stones and rocks.
NOTES	The world's smallest pine. A member of the podocarp family, the Podocarpaceae.

Large-flowered
Hebe

RM

Ruapehu Hebe

CM

Pygmy Pine

RM

Weeping Matipo/Mapou   *Myrsine divaricata*

CHARACTERISTICS	Height to 3 m.
BIOLOGY	Flowers from June to November. Fruits from August to May.
DISTRIBUTION	Throughout New Zealand and the Subantarctic Islands from coastal areas to the alpine zone.
HABITAT	Damp forests, shrublands and fellfields. Common in montane totara forests.
NOTES	A member of the myrsine family, the Myrsinaceae.

Hooker's Parahebe   *Parahebe hookeriana*

CHARACTERISTICS	Height to 60 cm.
BIOLOGY	Flowers from December to March.
DISTRIBUTION	In subalpine and alpine areas of the central North Island.
HABITAT	Along streambanks and among and emerging from under rocks.
NOTES	A member of the figwort family, the Scrophulariaceae.

Little Mountain Heath   *Pentachondra pumila*

CHARACTERISTICS	Forms mats up to 40 cm across.
BIOLOGY	Flowers from November to February. Fruits from November to April.
DISTRIBUTION	In subalpine and alpine areas throughout New Zealand.
HABITAT	Rocky areas, tussocklands, fellfields as well as along the margins of bogs.
NOTES	A member of the southern heath family, the Epacridaceae.

Weeping
Matipo

JB

Hooker's
Parahebe

CM

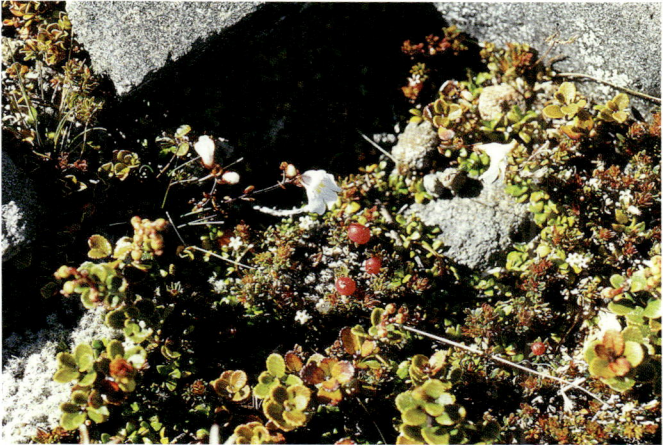

Little Mountain
Heath

JB

Herbs

One of the most attractive and diverse plant assemblages to be found in New Zealand is that presented by our alpine flora.

New Zealand is home to about 600 species of alpine plants, of which 93 percent are endemic. This includes eight entirely endemic genera and a further fifteen genera which are predominantly found here. The speargrasses, *Aciphylla* spp., for example, have some 40 species so far recognised, and apart from a single species in Australia and one on the Chathams, are entirely confined to mainland New Zealand.

One feature of New Zealand alpine plants that immediately strikes the visitor is that many have flowers that are white or yellow, and that the blues, reds and oranges so characteristic of alpine regions elsewhere are seldom seen.

The reason for this is that most of the pollinating insects in New Zealand are insects like flies, moths, beetles and certain native bees. These insects are short-tongued and specialise in short-tubed flowers that tend to be white or yellow. Butterflies and long-tongued bee species that respond better to more vibrant colours are relatively rare in New Zealand.

Blue Mountain Bidibidi/Piripiri   *Acaena inermis*

CHARACTERISTICS	Size to 75 cm across.
BIOLOGY	Flowers from December to February.
DISTRIBUTION	Commonly in montane and subalpine zones of the South Island, less commonly in alpine areas.
HABITAT	In short grasslands, on stony ground and along margins of screes.
NOTES	A member of the rose family, the Rosaceae.

Red Bidibidi/Piripiri   *Acaena microphylla*

CHARACTERISTICS	Size to 75 cm across.
BIOLOGY	Flowers from December to February.
DISTRIBUTION	In montane and subalpine zones of the Volcanic Plateau of the North Island, but commonest in the Tongariro National Park.
HABITAT	In tussocklands and rocky areas and on river flats.

Scarlet Bidibidi/Piripiri   *Acaena novaezelandiae*

CHARACTERISTICS	Forms mats to 75 cm across.
BIOLOGY	Flowers from September to December.
DISTRIBUTION	Throughout New Zealand from lowland to alpine areas.
HABITAT	In grasslands and shrublands and on roadsides, with a preference for the more open, drier areas.

Blue Mountain Bidibidi

CM

Red Bidibidi

RM

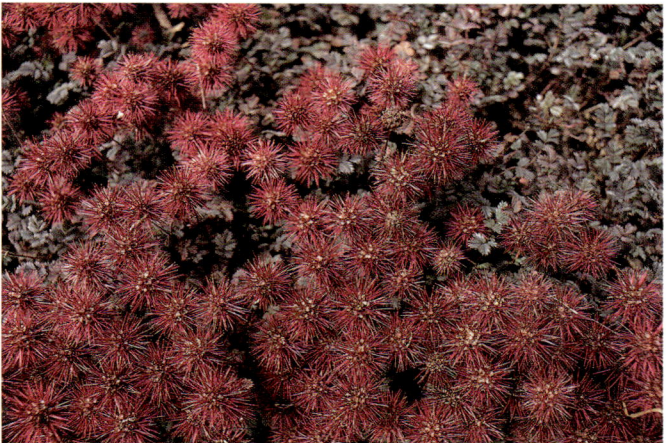

Scarlet Bidibidi

RM

Golden Spaniard *Aciphylla aurea*

CHARACTERISTICS	Height to 1 m.
BIOLOGY	Flowers in December and January.
DISTRIBUTION	From Nelson and Marlborough to Southland in montane to low-alpine areas on the eastern side of the Main Divide.
HABITAT	In tussocklands, shrubland and open scrub.
NOTES	A member of the carrot family, the Umbelliferae.

Wild Spaniard/Taramea *Aciphylla colensoi*

CHARACTERISTICS	Height to 1 m.
BIOLOGY	Flowers in November and December.
DISTRIBUTION	Widespread from Mount Hikurangi in the North Island south to about Arthur's Pass in the South Island in subalpine and low-alpine areas.
HABITAT	In tussocklands, shrublands and open scrub.

Snowball Spaniard *Aciphylla congesta*

CHARACTERISTICS	Height to 60 cm.
BIOLOGY	Flowers in January.
DISTRIBUTION	Subalpine areas from South Westland through to West Otago and Fiordland. Somewhat sporadic in distribution.
HABITAT	In damper areas of fellfields and rocky basins, as well as along streambanks and along the margins of tarns.

Golden Spaniard

RM

Wild Spaniard

CM

Snowball Spaniard

NS

Horrid Spaniard *Aciphylla horrida*

CHARACTERISTICS	Height to 1 m.
BIOLOGY	Flowers in December and January.
DISTRIBUTION	In higher-rainfall areas of the South Island to the west of the Main Divide from Arthur's Pass south to Fiordland from montane to alpine areas.
HABITAT	Open scrub and tussocklands.

Subalpine Spaniard *Aciphylla pinnatifida*

CHARACTERISTICS	Height to 20 cm.
BIOLOGY	Flowers in December and January.
DISTRIBUTION	Somewhat local in distribution in subalpine areas from Fiordland through to the Umbrella Mountains in West Otago.
HABITAT	Tussocklands, fellfields and open scrubland with a preference for damper areas.

Giant Spaniard *Aciphylla scott-thomsonii*

CHARACTERISTICS	Height to 1 m.
BIOLOGY	Flowers in December and January.
DISTRIBUTION	In subalpine areas of the South Island from Mount Cook southwards, but commoner in the wetter areas to the west of the Main Divide.
HABITAT	Tussocklands, fellfields and open scrub.

Horrid
Spaniard

RM

Subalpine
Spaniard

NS

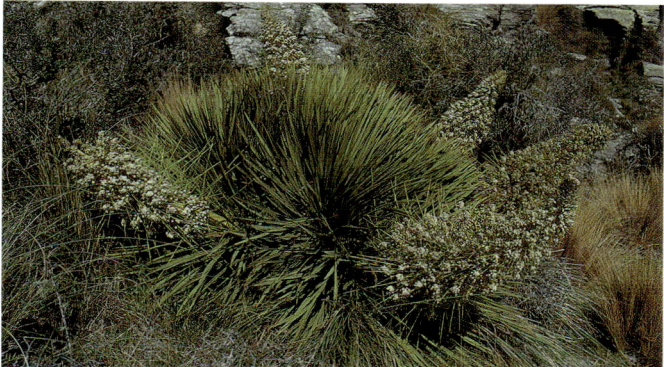

Giant
Spaniard

RM

Otago Spaniard   *Aciphylla simplex*

CHARACTERISTICS	Height to 30 cm.
BIOLOGY	Flowers in December and January.
DISTRIBUTION	Subalpine and low-alpine mountain areas in West Otago. One of the most restricted in range of the Spaniards.
HABITAT	In grasslands and cushionfields.

Feathery Spaniard/Taramea   *Aciphylla squarrosa*

CHARACTERISTICS	Height to 1 m.
BIOLOGY	Flowers in December and January.
DISTRIBUTION	In subalpine areas of the central North Island, from Mount Hikurangi southwards, descending to the coastline near Wellington.
HABITAT	Tussocklands, fellfields and open scrub.

Kopoti   *Anisotome aromatica*

CHARACTERISTICS	Height to 50 cm.
BIOLOGY	Flowers from October to December.
DISTRIBUTION	In subalpine to low-alpine areas throughout the North, South and Stewart islands.
HABITAT	In grasslands, rocky areas, tussock-herbfields, fellfields and areas of gravel and scoria.
NOTES	A member of the carrot family, the Umbelliferae.

Otago Spaniard

NS

Feathery Spaniard

JB

Kopoti

JB

Bristly Carrot   *Anisotome pilifera*

CHARACTERISTICS	Height to 60 cm.
BIOLOGY	Flowers from November to March.
DISTRIBUTION	Subalpine to nival zones in the mountains of the South Island from northwest Nelson southwards but absent from much of Central Otago.
HABITAT	Rocky areas.
NOTES	The long, strong taproot of this plant enables it to survive in situations hostile to other plants.

Mountain Astelia/Kakaha   *Astelia nervosa*

CHARACTERISTICS	Height to 80 cm.
BIOLOGY	Flowers from November to January. Fruits from March to May.
DISTRIBUTION	Montane and subalpine areas throughout the New Zealand mainland.
HABITAT	Moist areas in grasslands and shrublands. Sometimes along bush margins.
NOTES	A member of the lily family, the Liliaceae.

Bog Lily   *Bulbinella gibbsii*

CHARACTERISTICS	Height to 60 cm.
BIOLOGY	Flowers from October to January.
DISTRIBUTION	In montane and low-alpine areas from the Tararua and Ruahine ranges of the North Island south to Stewart Island, but commoner in areas of higher rainfall.
HABITAT	Moist areas on river flats and in tussocklands and herbfields.
NOTES	A member of the lily family, the Liliaceae.

Bristly Carrot

NS

Mountain
Astelia

CM

Bog Lily

RM

Maori Onion *Bulbinella hookeri*

CHARACTERISTICS	Height to 60 cm.
BIOLOGY	Flowers from October to January.
DISTRIBUTION	In montane areas to low-alpine areas from the Volcanic Plateau in the North Island south to North Canterbury.
HABITAT	Moist sites on river flats and in tussocklands and herbfields.

Mountain Bog Daisy *Celmisia alpina*

CHARACTERISTICS	Height to 30 mm.
BIOLOGY	Flowers from October to January.
DISTRIBUTION	Widespread in montane to low-alpine areas throughout the South Island.
HABITAT	Margins of tarns and bogs and in swampy places.
NOTES	A member of the daisy family, the Asteraceae.

Downy Daisy *Celmisia glandulosa*

CHARACTERISTICS	Height to 25 mm.
BIOLOGY	Flowers in December and January.
DISTRIBUTION	In subalpine and low-alpine areas from the Volcanic Plateau south to Fiordland, but uncommon in the drier areas east of the Main Divide.
HABITAT	Bogs and swampy margins of tarns. Also in permanently moist areas of tussocklands, tussock-herbfields and open scrub.

Maori Onion

CM

Mountain Bog
Daisy

NS

Downy Daisy

JB

Dainty Daisy/Pekapeka *Celmisia gracilenta*

CHARACTERISTICS	Height to 15 cm.
BIOLOGY	Flowers from November to February.
DISTRIBUTION	Throughout mainland New Zealand in montane, subalpine and alpine areas up to about 1600 m altitude.
HABITAT	In herbfields, grassfields and fellfields.

Mountain Daisy *Celmisia incana*

CHARACTERISTICS	Height to 40 mm.
BIOLOGY	Flowers December and January.
DISTRIBUTION	In subalpine and alpine areas of the North and South islands from the Coromandel Peninsula south to Otago.
HABITAT	In fellfields, herbfields, grassfields and exposed rocky areas.

False Spaniard *Celmisia lyallii*

CHARACTERISTICS	Height to 45 mm.
BIOLOGY	Flowers December and January.
DISTRIBUTION	In subalpine and low-alpine areas east of the Main Divide of the South Island from Marlborough to Southland.
HABITAT	In tussocklands, snow tussockfields and occasionally in open scrub and on sub-nival rock.

Dainty Daisy

Mountain Daisy

False Spaniard

Marlborough Daisy *Celmisia monroi*

CHARACTERISTICS	Height to 20 cm.
BIOLOGY	Flowers December and January.
DISTRIBUTION	Montane and subalpine areas of Nelson and Marlborough with a few records from the Canterbury Alps.
HABITAT	Rocky areas, snow tussocklands and tussock-herbfields.

Tikumu *Celmisia semicordata*

CHARACTERISTICS	Height to 60 cm.
BIOLOGY	Flowers December to February.
DISTRIBUTION	In montane and subalpine areas throughout the South Island from Nelson south to Fiordland.
HABITAT	Rocky areas, fellfields, snow tussocklands and tussock-herbfields.

Cushion Daisy *Celmisia sessiliflora*

CHARACTERISTICS	Height to 30 cm.
BIOLOGY	Flowers December and January.
DISTRIBUTION	Widespread in montane to alpine areas of the South and Stewart islands but sometimes fairly localised in distribution.
HABITAT	In tussocklands, tussock-herbfields, cushion bogs and open shrublands. Commoner in damper areas.

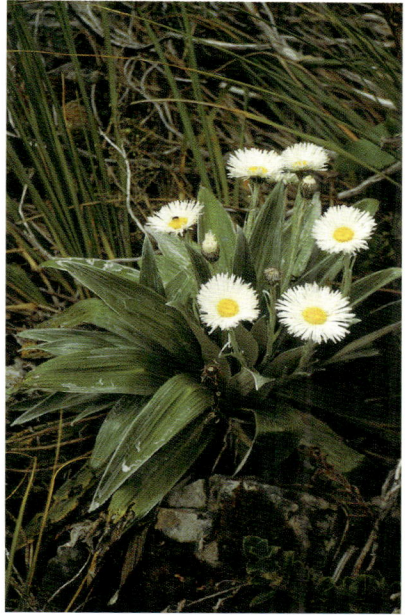

ABOVE LEFT: Marlborough Daisy

ABOVE RIGHT: Tikumu

RIGHT: Cushion Daisy

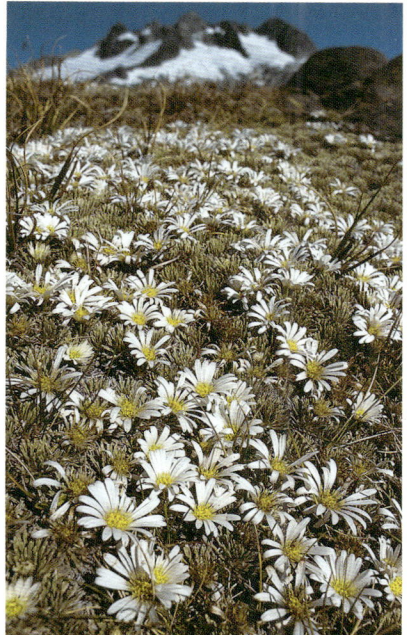

Cotton Daisy/Puakaito *Celmisia spectabilis*

CHARACTERISTICS	Height to 15 cm.
BIOLOGY	Flowers December and January.
DISTRIBUTION	In montane and low-alpine areas from the East Cape of the North Island south to North Otago.
HABITAT	In herbfields, tussocklands and fellfields.

Fiordland Mountain Daisy *Celmisia verbascifolia*

CHARACTERISTICS	Height to 40 cm.
BIOLOGY	Flowers January and February.
DISTRIBUTION	Montane and subalpine areas of the South Island from Nelson south to Fiordland, with a preference for areas of higher rainfall.
HABITAT	Tussock-herbfields, fellfields and rocky areas.

Woollyhead/Puatea *Craspedia uniflora*

CHARACTERISTICS	Height to 12 cm.
BIOLOGY	Flowers January and February. Fruits January to March.
DISTRIBUTION	From coastal to alpine areas throughout New Zealand south of the East Cape but reaching lower altitudes more commonly in the south.
HABITAT	Herbfields, streambeds, shrubland and wetlands with a preference for damper sites.
NOTES	A member of the daisy family, the Asteraceae.

Cotton Daisy

Fiordland Mountain Daisy

Woollyhead

Yellow Snow Marguerite *Dolichoglottis lyallii*

CHARACTERISTICS | Height to 50 cm.

BIOLOGY | Flowers December and January. Fruits January and February.

DISTRIBUTION | From montane to high-alpine areas of the South and Stewart islands but fairly localised in drier areas.

HABITAT | Permanently damp but well-lit places on herbfields, tussock-herbfields, tussocklands, rocky areas and fellfields.

NOTES | A member of the daisy family, the Asteraceae.

Snow Marguerite *Dolichoglottis scorzoneroides*

CHARACTERISTICS | Height to 60 cm.

BIOLOGY | Flowers December to February. Fruits February to April.

DISTRIBUTION | From montane to high-alpine areas of the South and Stewart islands. Widespread in higher-rainfall areas such as those to the west of the Main Divide.

HABITAT | Permanently damp but well-lit places on herbfields, tussock-herbfields, tussocklands, rocky areas and fellfields.

Alpine Sundew/Wahu *Drosera arcturi*

CHARACTERISTICS | Height to 12 cm.

BIOLOGY | Flowers from November to February. Fruits in February.

DISTRIBUTION | In montane to low-alpine zones from the Volcanic Plateau in the central North Island south to Stewart Island.

HABITAT | In bogs and swamps and moister areas around tarns and alpine lakes.

NOTES | An insectivorous species. A member of the sundew family, the Droseraceae.

CM

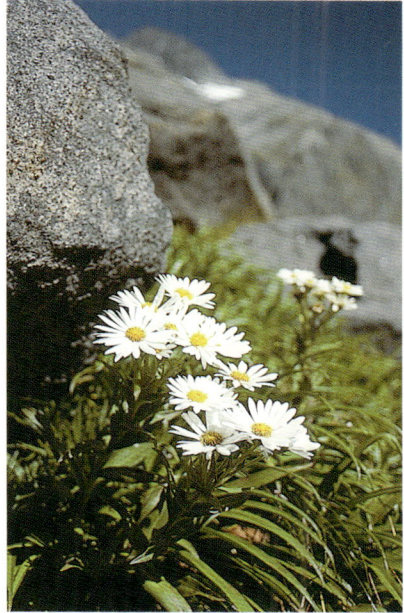

RM

ABOVE LEFT: Yellow Snow Marguerite

ABOVE RIGHT: Snow Marguerite

RIGHT: Alpine Sundew

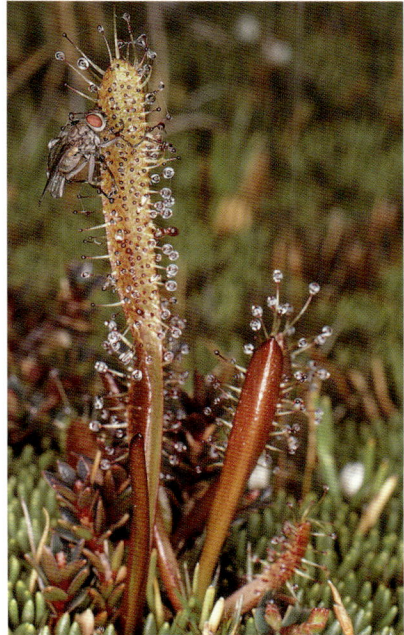

RM

Willowherb *Epilobium crassum*

CHARACTERISTICS	Size to 30 cm across.
BIOLOGY	Flowers in January. Fruits in February.
DISTRIBUTION	Low- to high-alpine areas of the South Island, from Nelson and Marlborough southwards to Southland.
HABITAT	Drier greywacke mountains on screes and fellfields.
NOTES	A member of the evening primrose family, the Onagraceae.

Scree Epilobium *Epilobium pychnostachyum*

CHARACTERISTICS	Size to 30 cm across.
BIOLOGY	Flowers in January, fruits in February.
DISTRIBUTION	Low- to high-alpine areas of the North and South islands from the Ruahine Ranges south to the Hawkdun Range in North Otago.
HABITAT	Drier greywacke mountains on screes and fellfields.

New Zealand Eyebright/Tutae-kiore *Euphrasia cuneata*

CHARACTERISTICS	Height to 60 cm.
BIOLOGY	Flowers from January to March. Fruits from February to March.
DISTRIBUTION	In montane and subalpine areas throughout the North Island and in the South Island south to North Canterbury.
HABITAT	Damp areas in fellfields, alpine herbfields and in subalpine scrub.
NOTES	A member of the figwort family, the Scrophulariaceae.

Willowherb

JB

Scree
Epilobium

CM

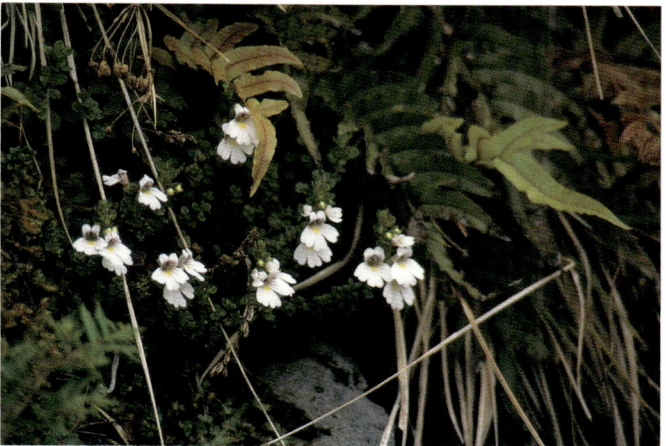

New Zealand
Eyebright

CM

Common New Zealand Gentian *Gentiana bellidifolia*

CHARACTERISTICS	Height to 15 cm.
BIOLOGY	Flowers from February to March.
DISTRIBUTION	From subalpine to high-alpine zones from Mount Hikurangi in the North Island south to Fiordland. Sometimes fairly localised in occurrence.
HABITAT	In fellfields, tussocklands, tussock-herbfields and herbfields.
NOTES	A member of the gentian family, the Gentianaceae.

Alpine Gentian *Gentiana patula*

CHARACTERISTICS	Height to 10 cm.
BIOLOGY	Flowers in January and February.
DISTRIBUTION	Montane to low-alpine zones from the Tararua Ranges in the North Island south to Southland, but rarely found in the wetter areas to the west of the Main Divide.
HABITAT	Common on moist sites in grasslands and tussocklands.

New Zealand Angelica/Naupiro *Gingidia montana*

CHARACTERISTICS	Height to 40 cm.
BIOLOGY	Flowers in November and December, fruits December to February.
DISTRIBUTION	Sporadic in distribution in lowland and to low-alpine zones of the North and South islands.
HABITAT	Moist, well-lit areas in tussocklands, tussock-herbfields and more open areas in scrub.
NOTES	Highly aromatic, smells of aniseed if leaves are crushed. A member of the carrot family, the Umbelliferae.

ABOVE LEFT: Common New Zealand
Gentian

ABOVE RIGHT: Alpine Gentian

RIGHT: New Zealand Angelica

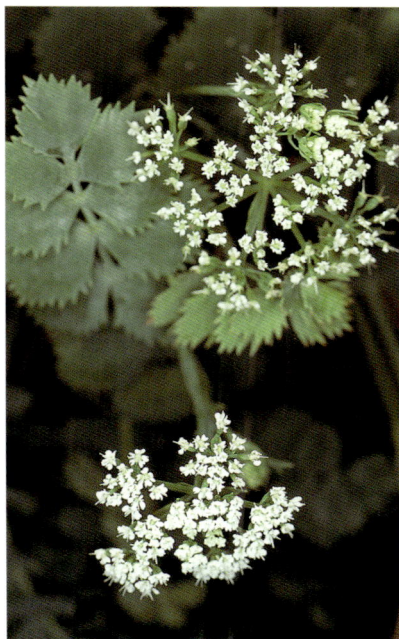

Giant Vegetable Sheep *Haastia pulvinaris*

CHARACTERISTICS	Up to 2 m across. Forms rounded or flat cushions.
BIOLOGY	Flowers January to March, fruits in March and April with the seeds being wind dispersed.
DISTRIBUTION	Low- to high-alpine areas in the mountains of Nelson and Marlborough.
HABITAT	Loose rocky areas above screes and in the more stable parts of screes. Also in fellfields.

Everlasting Daisy/Niniao *Helichrysum bellidioides*

CHARACTERISTICS	Height to 60 cm.
BIOLOGY	Flowers from November to February. Fruits in January and February.
DISTRIBUTION	Lowland to low-alpine zones from the East Cape of the North Island south to Stewart Island. Also on the Subantarctic Islands.
HABITAT	On roadsides, tussocklands, riverbeds, burned-over forests, rocky outcrops and herbfields.
NOTES	Frequently hybridises with another Everlasting Daisy, or Cudweed, *Gnaphalium hookeri*. A member of the daisy family, the Asteraceae.

Marlborough Helichrysum *Helichrysum coralloides*

CHARACTERISTICS	Height to 60 cm.
BIOLOGY	Flowers from November to February. Fruits in January and February.
DISTRIBUTION	Montane to high-alpine areas of the Seaward and Inland Kaikoura Ranges of the South Island.
HABITAT	Generally confined to rocky outcrops in drier, more open areas.

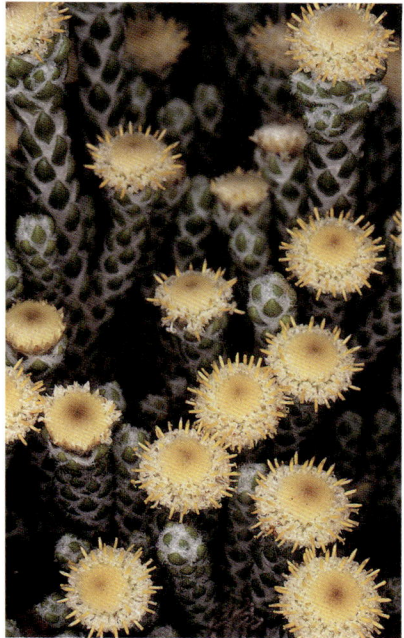

ABOVE LEFT: Giant Vegetable Sheep

ABOVE RIGHT: Everlasting Daisy

RIGHT: Marlborough Helichrysum

Yellow-flowered Helichrysum   *Helichrysum parvifolium*

CHARACTERISTICS	Height to 40 cm.
BIOLOGY	Flowers from December to March. Fruits in January and February.
DISTRIBUTION	Subalpine to high-alpine areas in the mountains of the South Island from northwest Nelson and Marlborough south to North Canterbury.
HABITAT	Crevices in rocky outcrops in drier, more exposed areas.

Yellow-flowered Daisy   *Kirkienella novaezelandiae*

CHARACTERISTICS	Forms cushions up to 1 m across. Flowers from December to February.
DISTRIBUTION	Alpine areas in the South Island from the Ben Ohau Range in Canterbury south to the Livingstone Range in Fiordland.
HABITAT	Rocky areas and screes.
NOTES	A member of the daisy family, the Asteraceae.

Kelleria Cushion   *Kelleria childii*

CHARACTERISTICS	Forms cushions up to 50 cm across.
BIOLOGY	Flowers from December to February.
DISTRIBUTION	High-alpine areas in the southern South Island from the Dunstan Range south to Lake Wakatipu.
HABITAT	Damper areas fringing streams.
NOTES	A member of the daphne family, the Thymelaeaceae.

Yellow-
flowered
Helichrysum

RM

Yellow-
flowered
Daisy

NS

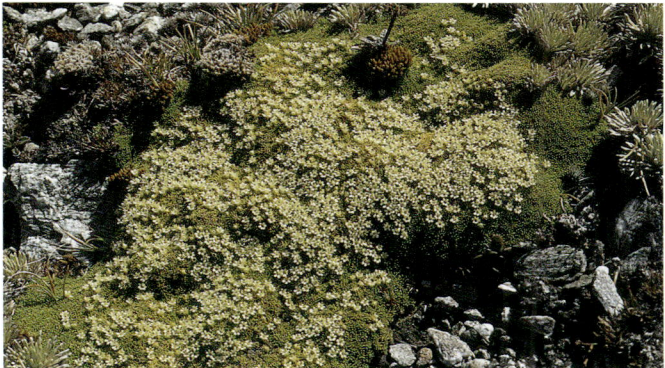

Kelleria
Cushion

NS

Black Daisy   *Leptinella atrata*

CHARACTERISTICS	Forms cushions to 40 cm across.
BIOLOGY	Flowers in January and February. Fruits in January and February.
DISTRIBUTION	From low- to high-alpine areas in the South Island from Marlborough to North Otago.
HABITAT	Screes on drier mountains.
NOTES	Formerly known as *Cotula atrata*. A member of the daisy family, the Asteraceae.

South Island Edelweiss   *Leucogenes grandiceps*

CHARACTERISTICS	Height 15–25 cm.
BIOLOGY	Flowers in January and February. Fruits in February and March.
DISTRIBUTION	Subalpine and alpine zones in the South and Stewart islands.
HABITAT	On rocky outcrops, in rocky alpine herbfields and fellfields, on rocky ledges and moraines.
NOTES	In the North Island this species is replaced by *Leucogenes leontopodium*. A member of the daisy family, the Asteraceae.

Grey-leaved Succulent   *Lignocarpa carnosula*

CHARACTERISTICS	Height to 15 cm.
BIOLOGY	Flowers from February to March.
DISTRIBUTION	Low- to high-alpine zones in the mountains of Canterbury and Marlborough in the South Island.
HABITAT	Among loose stones on scree slopes.
NOTES	Because of its cryptic colouration this plant is easily overlooked. A member of the carrot family, the Umbelliferae.

Black Daisy

South Island
Edelweiss

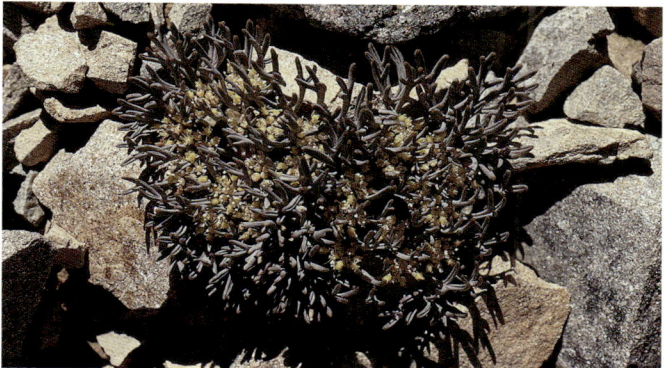

Grey-leaved
Succulent

Fleshy Lobelia *Lobelia roughii*

CHARACTERISTICS	Height to 10 cm.
BIOLOGY	Flowers from October to February. Fruits from December to April.
DISTRIBUTION	Low- to high-alpine areas of the South Island, from Nelson and Marlborough south to North Otago.
HABITAT	On shingle screes and rocky outcrops in drier areas. Especially prevalent in greywacke mountains.
NOTES	A member of the lobelia family, the Lobeliaceae.

Russel Lupin *Lupinus polyphyllus*

CHARACTERISTICS	Height to 60 cm.
BIOLOGY	Flowers from December to March.
DISTRIBUTION	Lowland to subalpine areas of the North, South and Stewart islands.
HABITAT	Disturbed and rough ground, forest margins, riverbeds, in secondary growth on roadside verges and stony areas.
NOTES	Particularly conspicuous in the Mackenzie country. An introduced species. A member of the pea family, the Papilionaceae.

Mountain Forget-me-not *Myosotis oreophila*

CHARACTERISTICS	Forms rosettes to 30 cm across.
BIOLOGY	Flowers from December to March.
DISTRIBUTION	Subalpine and alpine areas of the Dunstan Mountains of Central Otago in the South Island.
HABITAT	Among cushion vegetation in rocky areas.
NOTES	A member of the borage family, the Boraginaceae.

Fleshy Lobelia

Russel Lupin — purple flower.

Mountain Forget-me-not

Dwarf Forget-me-not *Myosotis pygmaea*

CHARACTERISTICS	Forms rosettes to 30 cm across.
BIOLOGY	Flowers from December to March.
DISTRIBUTION	Coastal to high-alpine zones of the South Island from Marlborough to South Canterbury east of the Main Divide.
HABITAT	Well-drained, open areas among depauperate vegetation in alpine zones in fellfields, snowbanks and on rocky outcrops.

Penwiper/Porotaka *Notothlaspi rosulatum*

CHARACTERISTICS	Height to 25 cm.
BIOLOGY	Flowers in December and January.
DISTRIBUTION	Subalpine to high-alpine zones of the South Island from Marlborough south to South Canterbury east of the Main Divide.
HABITAT	In the finer mixed debris of screes and the stony areas of fellfields and in the drier areas of greywacke mountains.
NOTES	A member of the mustard family, the Cruciferae.

North Island Mountain Foxglove/Hue-o-Raukatauri
Ourisia macrophylla

CHARACTERISTICS	Height to 10 cm.
BIOLOGY	Flowers from October to February. Fruits in February and March.
DISTRIBUTION	In subalpine and alpine areas of the North Island from the Volcanic Plateau south to the Tararua Ranges.
HABITAT	Damp, shady areas along streambanks and in herbfields.
NOTES	A member of the figwort family, the Scrophulariaceae.

Dwarf Forget-me-not

RM

Penwiper

RM

North Island Mountain Foxglove

JB

Matted Ourisia *Ourisia vulcanica*

CHARACTERISTICS	Forms mats 10–15 cm across.
BIOLOGY	Flowers from October to January. Fruits in February.
DISTRIBUTION	Subalpine and alpine zones of the Kaimanawa Ranges and the Volcanic Plateau in the North Island.
HABITAT	Exposed rocky areas in herbfields and fellfields.

Panakenake *Pratia angulata*

CHARACTERISTICS	Forms mats to 1 m across.
BIOLOGY	Flowers from October to March. Fruits in March.
DISTRIBUTION	From lowland to alpine zones throughout mainland New Zealand.
HABITAT	Damp and sheltered places in grassland, shrublands, scrub and in more open forests. Also on sand dunes and in rock crevices in coastal areas.
NOTES	A member of the lobelia family, the Lobeliaceae.

Large White-flowered Buttercup *Ranunculus buchananii*

CHARACTERISTICS	Height to 20 cm.
BIOLOGY	Flowers in December and January. Fruits in February and March.
DISTRIBUTION	More southerly, high-alpine areas of the South Island from South Westland to Fiordland and Otago and Southland.
HABITAT	Damp, rocky clefts, fellfields and especially in cirques.
NOTES	Because of its tolerance for cold, this is among our highest flowering plants. A member of the buttercup family, the Ranunculaceae.

Matted
Ourisia

CM

Panakenake

RM

Large White-
flowered
Buttercup

RM

Small Feathery-leaved Buttercup *Ranunculus gracilipes*

CHARACTERISTICS	Height to 15 cm.
BIOLOGY	Flowers in November and December. Fruits in January.
DISTRIBUTION	Sporadic distribution in the South Island from Nelson-Marlborough to Fiordland and on Stewart Island in subalpine to high-alpine zones but absent from the Kaikoura Ranges.
HABITAT	In scrub, shrubland, tussocklands and in sheltered rocky areas.

Haast's Buttercup *Ranunculus haastii*

CHARACTERISTICS	Height to 15 cm.
BIOLOGY	Flowers from November to January. Fruits in February.
DISTRIBUTION	Low- to high-alpine areas subject to high erosion in the greywacke mountains of the South Island from the Seaward Kaikoura Range south to the Takitimu Mountains.
HABITAT	Screes and slopes of shifting debris.

Korikori *Ranunculus insignis*

CHARACTERISTICS	Height to 90 cm.
BIOLOGY	Flowers from November to January. Fruits in February.
DISTRIBUTION	Subalpine to low-alpine areas of the North and South islands from the East Cape south to the Kaikoura Ranges.
HABITAT	Prefers areas of high rainfall where it grows in shrublands, snow-tussock herbfields, on streambanks and in damp clefts, flushes and hollows and on sheltered bluffs.

Above left: Small Feathery-leaved Buttercup

Above right: Haast's Buttercup

Right: Korikori

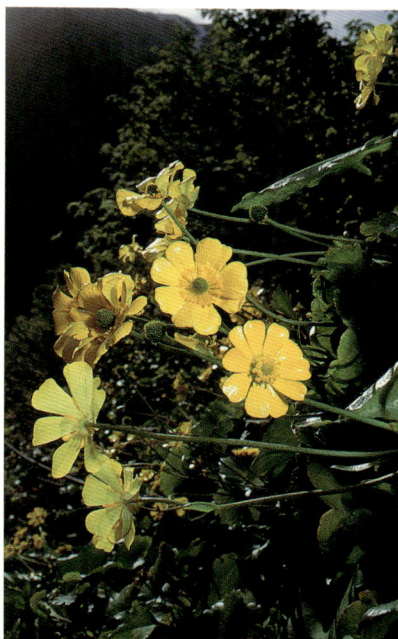

Mount Cook Lily/Kopukapuka *Ranunculus lyallii*

CHARACTERISTICS	Height to 1.5 cm.
BIOLOGY	Flowers from October to January. Fruits from December to March.
DISTRIBUTION	Subalpine to low-alpine areas of the South and Stewart islands.
HABITAT	Prefers areas of higher rainfall where it grows in shrublands, snow-tussock herbfields, wet hollows and flushes and along streambanks.
NOTES	Eliminated by browsing deer and chamois from many parts of its range. Although often erroneously called a lily this plant is actually a buttercup.

Snow Buttercup/Kawariki *Ranunculus nivicolus*

CHARACTERISTICS	Height to 80 cm.
BIOLOGY	Flowers from December to February. Fruits in March.
DISTRIBUTION	To about 1800 m in the Kaweka and Raukumara ranges, Mount Egmont/Taranaki and the Volcanic Plateau of the North Island.
HABITAT	Scrublands, fellfields and scoria slopes.

Lobe-leaf Buttercup *Ranunculus verticillatus*

CHARACTERISTICS	Height to 20 cm.
BIOLOGY	Flowers in December and January. Fruits in March.
DISTRIBUTION	From Mount Hauhungatahi in the central North Island through the Tararua and Ruahine ranges south to Nelson and Marlborough.
HABITAT	Permanently damp situations among rocks, in scrub, shrublands and fellfields.

Mount Cook
Lily

RM

Snow
Buttercup

CM

Lobe-leaf
Buttercup

JB

Volcanic Plateau Raoulia/Tutahuna   *Raoulia albo-sericea*

CHARACTERISTICS | Forms mats 60–80 cm across.

BIOLOGY | Flowers in December and January. Fruits in February.

DISTRIBUTION | Subalpine areas of the Volcanic Plateau in the North Island.

HABITAT | Depauperate stony areas, fellfields and areas of pumice.

NOTES | A member of the daisy family, the Asteraceae.

Scabweed   *Raoulia australis*

CHARACTERISTICS | Forms mats 30–50 cm across.

BIOLOGY | Flowers in January. Fruits in February.

DISTRIBUTION | Subalpine areas in the mountains of the South Island from Arthur's Pass south to the Hollyford Valley in Fiordland.

HABITAT | In open grass and tussocklands, on stable river flats and in fellfields.

Common Vegetable Sheep/Tutahuna   *Raoulia eximia*

CHARACTERISTICS | Forms mats 20–60 cm across.

BIOLOGY | Flowers in January. Fruits from January to March.

DISTRIBUTION | Low- to high-alpine zones of the South Island from Nelson and Marlborough south to North Otago.

HABITAT | Dry rocky ground and more stable areas in fellfields.

Volcanic
Plateau
Raoulia

CM

Scabweed

RM

Common
Vegetable
Sheep

RM

Mat Daisy *Raoulia hookeri*

CHARACTERISTICS	Forms mats from 30–50 cm across.
BIOLOGY	Flowers in January. Fruits in February.
DISTRIBUTION	Low- to high-alpine areas from the Volcanic Plateau in the North Island south to Nelson.
HABITAT	Stony riverbeds, fellfields and depauperate sites in tussocklands.

Lawyer/Tataramoa *Rubus schmidelioides*

CHARACTERISTICS	In open areas forms mounds several metres across. Flowers from September to December. Fruits from February to April.
DISTRIBUTION	Lowland to subalpine areas of the North, South and Stewart islands.
HABITAT	A prickly climber in forest on valley floors and river flats. Less commonly found in scrub and shrublands.
NOTES	A member of the rose family, the Rosaceae.

Mountain Chickweed/Kohukohu *Stellaria roughii*

CHARACTERISTICS	Forms mats 20–40 cm across.
BIOLOGY	Flowers from November to February.
DISTRIBUTION	Low- to high-alpine areas subject to high erosion in the greywacke mountains of the South Island, from Nelson south to the Takitimu Mountains of western Southland.
HABITAT	Screes and slopes of shifting debris.
NOTES	A member of the chickweed family, the Caryophyllaceae.

Mat Daisy

RM

Lawyer

JB

Mountain
Chickweed

RM

New Zealand Bluebell   *Wahlenbergia albomarginata*

CHARACTERISTICS	Height to 25 cm.
BIOLOGY	Flowers from December to March.
DISTRIBUTION	Lowland to low-alpine areas of the South and Stewart islands.
HABITAT	Herbfields, fellfields and tussock grasslands and occasionally on scree with a preference for well-drained sites.
NOTES	Also sometimes called a harebell. Belongs to the bellflower family, the Campanulaceae.

Maori Bluebell/Rimuroa   *Wahlenbergia pygmaea*

CHARACTERISTICS	Height to 8–10 cm.
BIOLOGY	Flowers from November to February.
DISTRIBUTION	Lowland to low-alpine areas from the Volcanic Plateau of the North Island south to Southland.
HABITAT	Herbfields, fellfields and tussock grasslands and valley floors in higher-rainfall areas, but with a preference for well-drained sites.

New Zealand Bluebell

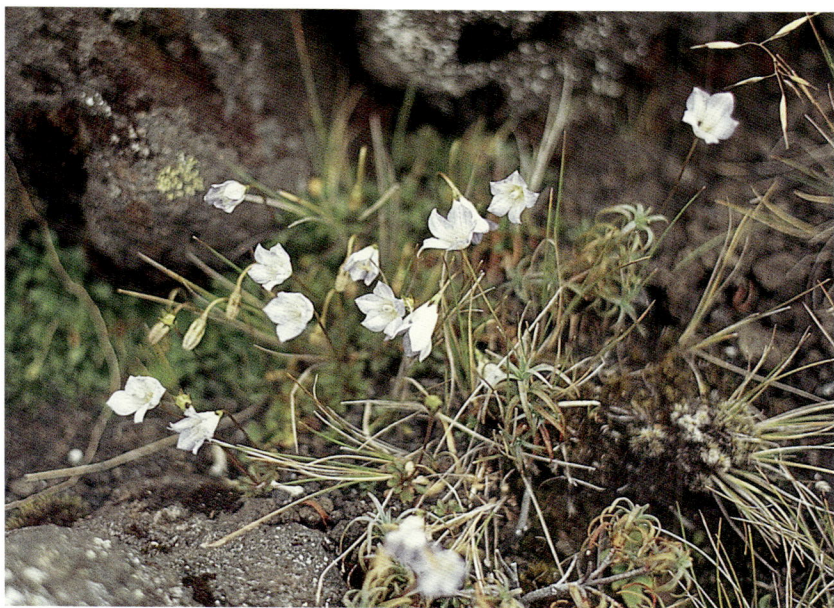

Maori Bluebell

Orchids

The orchids are among the largest of all plant families, with some 20,000 species occurring worldwide. There are both terrestrial and epiphytic types.

Orchids produce some of the most spectacular of all flowers in nature. Although the New Zealand examples are never as spectacular as their tropical cousins, they are nevertheless among the most colourful members of our alpine flora.

Orchids are also among the few native plants with blue flowers. The Blue Sun Orchid is one of the most spectacular of our orchids, as well as being one of the most widespread. It occurs from about Rotorua south to Port Pegasus on Stewart Island, from the coast up into montane areas.

The edible tubers of several native orchid species were traditionally harvested by Maori.

Odd-leaved Orchid   *Aporostylis bifolia*

CHARACTERISTICS	Height to 20 cm.
BIOLOGY	Flowers in December and January.
DISTRIBUTION	From lowland to subalpine areas throughout mainland New Zealand. Also on the Chathams and some subantarctic islands.
HABITAT	Damp grasslands and herbfields as well as on the margins of tarns and bogs.
NOTES	A member of the orchid family, the Orchidaceae.

White Mountain Orchid   *Caladenia lyalli*

CHARACTERISTICS	Height 10–30 cm.
BIOLOGY	Flowers December and January.
DISTRIBUTION	Montane and subalpine areas of the North, South and Stewart islands, but local and uncommon in occurrence.
HABITAT	Damp and boggy areas in herbfields, tussock-herbfields, shrublands and in light scrub.

Black Orchid/Huperei   *Gastrodia cunninghamii*

CHARACTERISTICS	Height to 1 m.
BIOLOGY	Flowers in January and February.
DISTRIBUTION	From lowland to montane zones throughout mainland New Zealand. Also on the Chathams.
HABITAT	Under native and exotic trees and in shrublands. Relatively common in beech forests.

Odd-leaved Orchid

White Mountain Orchid

Black Orchid

Tutukiwi *Pterostylis patens*

CHARACTERISTICS	Height to 50 cm.
BIOLOGY	Flowers from September to December.
DISTRIBUTION	Coastal to montane areas of mainland New Zealand. Also on the Chathams.
HABITAT	Damp and shady places in shrublands, forest and also not uncommon along forest edges.

Blue Sun Orchid *Thelymitra hatchii*

CHARACTERISTICS	Height 10–45 cm.
BIOLOGY	Flowers in January and February.
DISTRIBUTION	Coastal to montane areas of the North, South and Stewart islands but fairly local in occurrence.
HABITAT	Damp and shady places in shrublands, in forests and also not uncommon along forest edges.

White Sun Orchid/Maika *Thelymitra longifolia*

CHARACTERISTICS	Height 10–45 cm.
BIOLOGY	Flowers November and December. Fruits January and February.
DISTRIBUTION	From coastal to montane areas from the Poor Knight Islands south to Stewart Island. Also on the Chatham and Auckland islands.
HABITAT	Tussocklands and tussock-herbfields.
NOTES	Flowers may be pink, blue or white.

Tutukiwi

Blue Sun Orchid

White Sun Orchid

Grasses

When European settlers first arrived in New Zealand they encountered a country covered in great swathes of tussock grasses and many of these early settlers thought that this was a natural occurrence. Recently, however, it has been found that these great stretches of grasslands were a relatively recent phenomenon resulting mainly from uncontrolled fires started by the early Maori.

There are definite habitat preferences among the grasses with the drier, low-altitude grasslands supporting short tussocks of the *Festuca* and *Poa* genera along with several local representatives of the *Rytidosperma*, a predominantly Australian genus.

In alpine herbfields and grassland tall tussocks prevail, among the most widespread of which are *Chionochloa macra*, *C. pallens* and *C. rigida* in the South Island and *C. rubra* in the North Island.

Our native grasses have had a great surge in popularity with gardeners in recent years, with new native species coming onto the market all the time.

Leather-leaf Carex/Pukio *Carex buchananii*

CHARACTERISTICS	Height to 30 cm.
BIOLOGY	Flowers in November and December. Fruits in January.
DISTRIBUTION	Thoughout mainland New Zealand in subalpine and alpine areas but somewhat localised in distribution.
HABITAT	Tussocklands and tussock-herbfields.
NOTES	A member of the sedge family, the Cyperaceae.

Mountain Sedge *Carpha alpina*

CHARACTERISTICS	Height to 30 cm.
BIOLOGY	Flowers in November and December. Fruits in February.
DISTRIBUTION	North, South and Stewart islands from coast to subalpine and alpine areas, but localised in occurrence. Also in Australia, Tasmania and New Guinea.
HABITAT	In and around bogs, swamps, tarns and wet ground.
NOTES	A member of the sedge family, the Cyperaceae.

Hunangamoho *Chionochloa conspicua*

CHARACTERISTICS	Height to 2 m. Among the tallest of the native grass species.
BIOLOGY	Flowers from December to February. Fruits from February to April.
DISTRIBUTION	Throughout New Zealand from coastal to alpine areas.
HABITAT	Streamsides, natural forest clearings, lowland and subalpine grasslands.
NOTES	Popular in cultivation. A member of the grass family, the Gramineae.

Leather-leaf
Carex

CM

Mountain
Sedge

JB

Hunangamoho

CM

Broad-leaved Snow Tussock/Haumata
Chionochloa flavescens

CHARACTERISTICS	Height to 60 cm.
BIOLOGY	Flowers from December to February. Fruits from February to April.
DISTRIBUTION	From the Tararua Ranges in the North Island south to Fiordland in subalpine to low-alpine areas.
HABITAT	Tussocklands, shrublands, short scrub and on cliff-faces.
NOTES	A related species, as yet undescribed, is found on Mount Anglem on Stewart Island.

Mid-ribbed Snow Tussock *Chionochloa pallens*

CHARACTERISTICS	Height to 2 m.
BIOLOGY	Flowers from December to February. Fruits from February to April.
DISTRIBUTION	From Mount Hikurangi and Mount Egmont/Taranaki in the North Island, southwards to Fiordland.
HABITAT	Tussocklands and tussock-herbfields up to 1700 m.
NOTES	This is the commonest South Island tussock.

Red Tussock *Chionochloa rubra*

CHARACTERISTICS	Height to 1.6 m.
BIOLOGY	Flowers in December and January. Fruits from January to March.
DISTRIBUTION	Localised in occurrence in the North, South and Stewart islands.
HABITAT	Subalpine and montane tussocklands and shrublands.
NOTES	Popular in cultivation.

Broad-leaved
Snow Tussock

CM

Mid-ribbed
Snow Tussock

RM

Copper
Tussock

RM

Tall Mountain Sedge/Maru *Gahnia procera*

CHARACTERISTICS | Height to 2 m.

BIOLOGY | Flowers from September to December. Fruits all year.

DISTRIBUTION | Mountains of Nelson, Marlborough and Westland.

HABITAT | Tussocklands and tussock-herbfields.

NOTES | A member of the sedge family, the Cyperaceae.

Bristle Tussock *Rytidosperma setifolium*

CHARACTERISTICS | Height to 30 cm.

BIOLOGY | Flowers from December to February. Fruits in February. Light brown in colour, with orange pollen.

DISTRIBUTION | In montane and subalpine areas throughout mainland New Zealand but sometimes fairly localised in occurrence.

HABITAT | Exposed stony places in meadows. Also in tussocklands and tussock-herbfields.

NOTES | A member of the grass family, the Gramineae.

Tall Mountain Sedge

Bristle Tussock

Ferns

Although some people think of ferns as tropical plants, ferns grow prolifically in a multitude of shapes and sizes in New Zealand, a country which is at times most definitely not tropical.

They cope with just about every environment imaginable, from dense forest to geothermal areas, and from storm-ravaged sea-cliffs to limestone crags. Although the vast majority of our ferns are lowland species, they are not uncommon in tussocklands, shrublands and fellfields, and wanderers in the mountains will consequently come across a number of species that are quite at home there.

Most ferns found in the alpine areas are upland forms of species that are also found elsewhere, unlike some such as the Alpine Shield Fern, one of our few deciduous ferns, which grows only in the mountains of the South Island with a satellite population on Mount Egmont/Taranaki.

Mountain Kiokio   *Blechnum montanum*

CHARACTERISTICS	Height to 45 cm.
BIOLOGY	Varies somewhat in appearance according to habitat, being smaller at higher altitudes.
DISTRIBUTION	Common in montane to subalpine zones throughout mainland New Zealand. Patchy distribution at other altitudes.
HABITAT	Open tussocklands, grasslands and herbfields. Also in light scrub, more open forest and along forest edges.
NOTES	Formerly incorrectly known as *Blechnum capense*. An easily cultivated fern. A member of the Blechnaceae family.

Alpine Hard Fern   *Blechnum penna-marina*

CHARACTERISTICS	Height to 15 cm.
BIOLOGY	Varies according to altitude, being smaller at higher altitudes.
DISTRIBUTION	Patchy distribution from coastal to subalpine zones throughout mainland New Zealand and the Subantarctic Islands.
HABITAT	Open tussocklands, grasslands and herbfields. Also in light scrub and along forest edges.

Alpine Shield Fern   *Polystichum cystostegia*

CHARACTERISTICS	Height to 15 cm.
DISTRIBUTION	Montane, subalpine and alpine zones of the South Island. Confined to Mount Egmont/Taranaki in the North Island.
HABITAT	In tussocklands, herbfields, screes, open rocky areas and on cliff-faces.
NOTES	An easily cultivated fern. Our highest-altitude fern, the Alpine Shield Fern is deciduous and dies off in winter, reappearing in the spring. A member of the Dryopteridaceae family.

Mountain
Kiokio

CM

Alpine Hard
Fern

RM

Alpine Shield
Fern

JB

Bracken/Rahurahu   *Pteridium esculentum*

CHARACTERISTICS	Height to 1.5 m. Sometimes forms large thickets.
BIOLOGY	Spores are dispersed in late spring and summer by wind and rain.
DISTRIBUTION	Throughout New Zealand from the Kermadecs to the Antipodes and from coastal to subalpine areas.
HABITAT	Open ground, shrublands and along forest edges. Colonises disturbed ground.
NOTES	Bracken roots were once harvested by Maori for food. A member of the Dennstaedtiaceae family.

Umbrella Fern /Tapuwae Kotuku
Sticherus cunninghamii

CHARACTERISTICS	Height to 50 cm. At times forms fairly large thickets.
BIOLOGY	Reproduces by spores and rhizomes.
DISTRIBUTION	From coastal to montane zones in the North Island, but generally at lower altitudes in the South Island and on Stewart Island.
HABITAT	Along roadside margins, on disturbed ground and in open areas.
NOTES	A very difficult fern to cultivate. From the Gleicheniaceae family.

Alpine Clubmoss   *Lycopodium fastigiatum*

CHARACTERISTICS	Height to 40 cm.
BIOLOGY	Produces stalked, erect, yellow-brown cones.
DISTRIBUTION	Throughout mainland New Zealand and the Subantarctic Islands.
HABITAT	Coastal to alpine grasslands, herbfields and boggy areas. Also in shrublands and among manuka.
NOTES	From the Lycopodiaceae family. Distinguished by cones and green sterile leaves that are sometimes orange in exposed conditions.

Bracken

RM

Umbrella Fern

CM

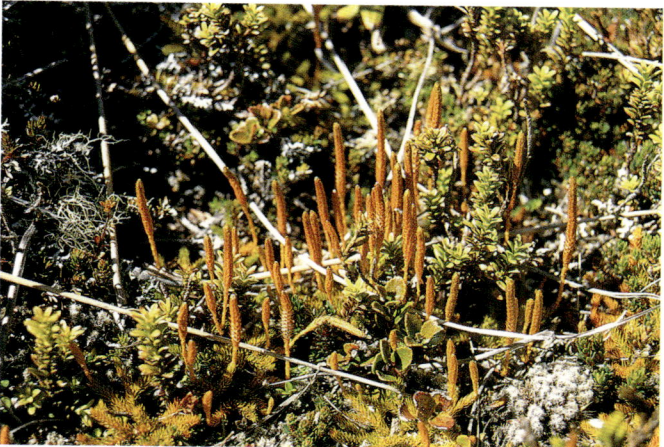

Alpine
Clubmoss

JB

Lower Plants

This group includes such plants as lichens, liverworts and mosses that are often overlooked by people interested in our biota. This is a pity as they are a fascinating and often beautiful group of plants. All are ancient plants that have been around since the Carboniferous period over 300 million years ago.

Lichens are an extremely diverse family with about 1000 species so far identified in New Zealand and probably as many yet to be classified and some of the largest species known are found here. They are remarkable in being able to survive in conditions that other 'higher' plants would find intolerable, living in places that often freeze over or in water that can reach boiling point. They are among the few plant species able to survive in the nival zone and they exist here in surprising numbers.

Lichens are very susceptible to pollution so the great diversity found in New Zealand is taken as an indication of our clean, green environment.

Mosses and liverworts are also primitive plants but despite the length of time they have been around are not fully adapted for living on land. The mosses of the alpine areas belong to a group known as the Andreae, which are mosses that have adapted to living in polar and alpine areas.

Lace Lichen *Cladia retipora*

CHARACTERISTICS	Forms extensive clumps of lace-like or coral-like branching growths that are either whitish or greyish in colour. Soft when wet.
BIOLOGY	Spores are spread by wind and rain.
DISTRIBUTION	Sporadic distribution from lowland to subalpine areas of the North, South and Stewart islands but less common at higher altitudes.
HABITAT	On dead vegetation and rocks. Sometimes on branches of living trees such as Manuka, *Leptospermum scoparium* and kKanuka, *Kunzea ericoides*.

Lecanora epibryon

CHARACTERISTICS	Forms white or greyish-white crusts 2–3 cm across with characteristic brownish-black discs.
BIOLOGY	Spores are spread by wind and rain.
DISTRIBUTION	Subalpine and alpine areas in the north of the South Island from Nelson south to Fiordland. Also somewhat locally distributed in subalpine areas of the lower North Island.
HABITAT	On the dead bases of tussocks.

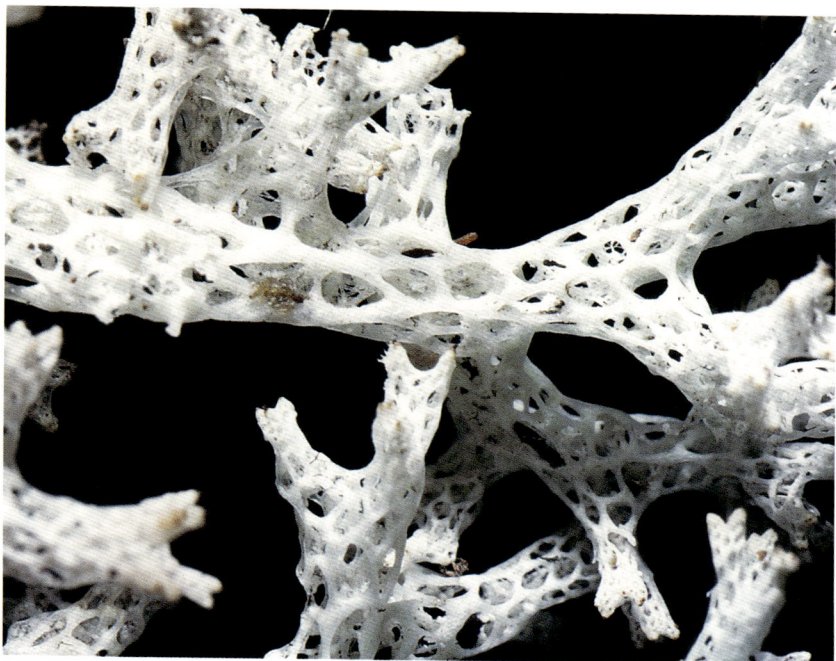

Lace Lichen

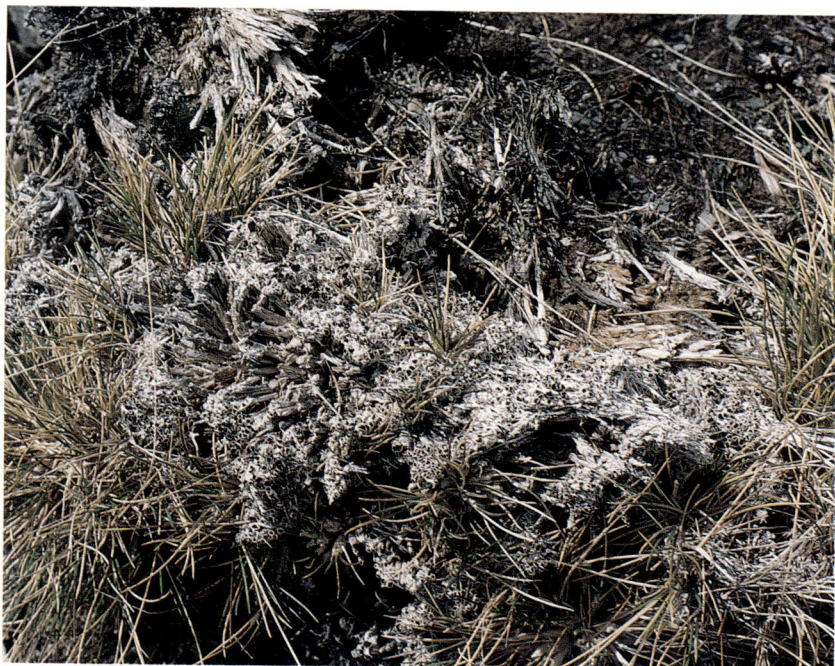

Lecanora epibryon

Sphagnum Moss/Kohukohu *Sphagnum cristatum*

CHARACTERISTICS	Pale green to yellowish-brown in colour.
BIOLOGY	Reproduces either asexually with pieces breaking off the parent plant and growing elsewhere, or sexually by spores that are 'blown out' of the the tips of branches.
DISTRIBUTION	From lowland to subalpine areas in the South Island but more common in the wetter areas to the west of the Main Divide. Also found on Stewart Island.
HABITAT	Forms soggy cushions in swamps, bogs and tarns and in damper areas in forests and shrublands.

Old Man's Beard/Angiangi *Usnea cilifera*

CHARACTERISTICS	Forms woolly, intricately branched clumps which range from greyish-white to light pinkish-white in colour.
BIOLOGY	Spores are dispersed by wind and rain.
DISTRIBUTION	From lowland to subalpine areas in the North, South and Stewart islands.
HABITAT	Drapes dead branches, shrubs and rocks.

RM

ABOVE:
Sphagnum Moss

RIGHT:
Old Man's
Beard

JB

Glossary

Aquatic phase	the period when an animal lives in water, usually in the juvenile stage.
Biota	all the animals and plants that are found in a particular area.
Chironomid larvae	the juvenile phase of species of the Midge family (Chironomidae).
Cryptic	hidden or well camouflaged.
Depauperate	impoverished or lacking in variety.
Endemic	plants and animals that are naturally restricted to a particular area.
Epiphytic	a non-parasitic plant that grows on another plant.
Fellfields	areas in the high-alpine zone with relatively little soil where plants such as Raoulia, Haastia and some of the the smaller Hebe species grow, often wedged in among rocks.
Feral	a domesticated animal, such as the cat, that has gone wild.
Form	a depression in vegetation in which hares live.
Herbfields	areas in the low-alpine zone dominated by herbs such as Celmisia, Ranunculus, Aciphylla, Astelia and Anisotome.
Herbivorus	eater of plant material.
Insectivorous	eater of insects.
Main Divide	the mountainous chain separating the east and west coasts of the South Island of New Zealand.
Mast years	the years, occurring at irregular intervals, when beech trees set seed.
Mustelids	members of the weasel family, the Mustelidae. Examples occurring in New Zealand are stoats, weasels and ferrets.
Nocturnal	active by night.
Orogeny	periods of intense, mountain-building activity.

Peneplain	a flat land surface caused by prolonged erosion in between orogenous periods.
Radula	the ribbon-like, rasping tongue found in snails and slugs.
Refugia	a protected area in which plants and animals survive, having been driven from their normal range by events such as intense glaciation.
Rhizomes	a thick, underground stem that produces aerial shoots.
Riffle	an area of fast-flowing and shallow water.
Rut	the period when mammals, such as deer, are 'on heat', or ready to mate.
Scree	areas of loose, unstable rock on uniformly steep slopes.
Simuliid larvae	the juvenile phase of species of the sandfly family (Simuliidae).
Solifluction	the downhill movement of rock as a result of periods of alternating heating and thawing.
spp.	the abbreviated plural form of the word species.
Stoner	a flightless stonefly.
Strobili	a cone or cone-like plant structure.
Stubble	areas of cropped vegetation.
Tarn	a small lake or large pond among mountains.
Tor	a rocky outcrop or pointed hill.

Index of common names

Fauna

Flora

Index of scientific names

Fauna

Flora